THE DAWN OF AN ASTONISHING ERA

Our world is by no means the center of the universe, but only one of billions of globes spinning in space. Although we are only beginning to gain experience of space/time, already some radio astronomers are turning our attention away to outer space, persuading us that we are not alone in the cosmos. In the multitude of solar systems, other civilizations have evolved, some of which have almost certainly mastered the techniques of radio electronics on the interstellar scale, and space travel.

The era of galactic communications is now opening up before us. . . . An era which will draw us closer to our un-earthly neighbors.

THE SEARCH FOR INTELLIGENT LIFE IN OUTER SPACE

Alfred Roulet
Translated and Edited by William A. Packer

A BERKLEY MEDALLION BOOK
published by
BERKLEY PUBLISHING CORPORATION

Published by arrangement with RENE JULLIARD Editeur

BERKLEY PUBLISHING CORPORATION
200 Madison Avenue
New York, N.Y. 10016

SBN 425-03319-8

BERKLEY MEDALLION BOOKS are published by
Berkley Publishing Corporation
200 Madison Avenue
New York, N.Y. 10016

BERKLEY MEDALLION BOOK ® TM 757,375

Printed in the United States of America

Berkley Medallion Edition, FEBRUARY, 1977

First published in Paris by RENE JULLIARD Editeur
under the title *A La Recherche des Extra-terrestres*

CONTENTS

PIONEERING NEW HORIZONS: THE UNIVERSE

On March 3, 1972, at Cape Kennedy in Florida, the United States Aeronautics and Space Administration shot another space engine into the sky. To the world at large, already accustomed to prodigious astronautical feats—the first Moon landings had occurred three years before—the event was hardly startling. Yet Pioneer 10, the latest and most advanced in a series of NASA space-exploring craft bearing the same name, was marked for a singular destiny; it would be the first electronical device ever conceived and built by human genius to traverse the solar system to its farthest confines and continue its flight into the infinite Galaxy beyond.

The swiftest of all vehicles hitherto produced by man, Pioneer 10 was assigned the immediate mission of surveying the planet Jupiter and relaying its findings back to Earth. Within eighteen hours it doubled the lunar orbit; after eight weeks in flight, it overshot the Martian trajectory. By the beginning of June, it was hurtling through the asteroid belt between Mars and Jupiter, a dangerous zone strewn with débris and fragments thought to be the remnants of some disintegrated planet. But fortunately encountering no mishaps, the spacecraft held firmly to its prearranged schedule and set its course for Jupiter, which it circled punctually in December, 1973, after racking up an astonishing flight log of nearly 600 million miles in twenty-one months.

Equipped with highly sophisticated electronic installations, the doughty little 120-pound vehicle duly performed its mission and transmitted remarkably clear television images of Jupiter back to Earth before heading into its escape trajectory. In 1984 or 1985 this true trail-blazing pioneer will traverse Pluto's orbit, most distant of the Earth's sister planets, approximately 3.5 billion miles from the Sun, and strike out for the cosmos.

By then, Pioneer 10 will have severed its contacts with the Earth. Once it has traveled a distance of two billion miles, all communications between its transmitting instruments and NASA headquarters will cease. The engine will then become a derelict mass of metal, racing through the Galaxy on a vagrant course, condemned to lose itself in interstellar space and wander through the Universe for perhaps millions of years. Should it eventually turn toward Taurus, some 1.7 million years will have elapsed before it reaches the environs of Aldebaran, brightest star in that constellation, approximately sixty-eight years from us.

Claiming still another primacy, Pioneer 10 bears the first message addressed by mankind to living beings in outer space. Although the chances are infinitesimal that it will ever be read, nevertheless the gesture opens up a fantastic new chapter in the annals of human adventure: man's quest to discover other beings endowed with intelligence.

Although the science of astronautics is fairly new, already it has influenced the people of the Earth to take a more realistic view of our position in the cosmos. Many have discovered suddenly, to their surprise, that our world is by no means the center of the Universe, but only one of billions of globes spinning in space. Although we are only beginning to gain experience of space-time, some radio astronomers are already turning our attention away to outer space, intent on imbuing us with a new galactic consciousness, and persuading us that we are not alone in the cosmos. In the multitude of solar systems, they tell us, other civilizations have evolved, some of which have almost certainly mastered the techniques of radio electronics on the interstellar scale, and perhaps of space travel.

In short, the era of galactic communications is now opening up before us.

For the present, this new adventure poses two immediate problems. First, in their effort to establish contacts with alien societies, assuming that intelligent life does indeed exist in outer space, teams of scientists are striving to mobilize the radio techniques. Secondly, it is no longer realistic to ignore arbitrarily the phenomenon of unidentified flying objects, those mysterious aerial visions observed in the Earth's atmosphere, which some scientists believe may well be associated with extra-terrestrials. The voluminous bulk of reports on UFO sightings amassed and examined will not likely dissipate our misgivings about their authenticity, yet various aspects of the phenomenon persistently defy critical analysis. The question has finally begun to preoccupy some qualified scientists: are UFOs actually products of outer space intelligence? A few astronomers and physicists, however averse they may be to any sort of occultism, are ready to admit such an eventuality.

Other scientists engaged in the study of outer space societies believe that UFOs should be assigned to an independent field of investigation, while still others dismiss them out of hand as a matter of no consequence. Yet, these fields, both championed by men dedicated to research into the possible, are in a certain sense complementary: on the one hand, a probabilistic, virtuoso exercise lacking any concrete verification, despite the hard logic of the assumptions they engender, on the other, a formidable accumulation of information and reports on UFO observations which, however incredible they may be, by no means sanction offhand, superficial interpretations. For it is "difficult to ignore the small residue of well-documented but unexplainable cases that form the hard-core of the UFO controversy"[1]

[1]American Institute of Astronautics and Aeronautics: UFO Symposium, 1970—An Appraisal of the Problem.

LIFE IN OUR GALACTIC HOMELAND

ONE OUT OF TWO HUNDRED BILLION STARS

In October, 1972, an impressive representation of the giants of the space world met at the Hofburg Palace in Vienna for a conference convoked by the International Astronautics Federation. The participants on hand—including a large American delegation headed by Dr. Wernher von Braun, and another from the Soviet Union, headed by Professor Leonid Sedov—were primarily concerned with the space programs then under way and their political, juridical and technical implications.

At the final session, however, this august assembly listened in sober silence as two of their members discussed a subject which would have struck an eavesdropping layman as wildly fantastic: civilizations living in outer space. Yet, without a single lift of an eyebrow anywhere in the hall, the listeners acknowledged the probability that societies not unlike our own do indeed inhabit other solar worlds in the Galaxy. The two speakers were by no means highly imaginative spinners of science fiction tales; on the contrary, one was a pragmatic astronomer, the other an engineer in advanced electronics.

The first, Sebastian Von Hoerner, a German-American astrophysicist and mathematician who had worked out a system for estimating the number of existing galactic societies and their evolution, expounded the latest developments in his research into the question. The second was Bernard M. Oliver, who discoursed on the newly-launched Project Cyclops which, under his direction, was trying to mobilize powerful radio-astronomic means of intercepting radio signals presumably transmitted by intelligent life from unknown points in the Galaxy. Since the earliest listening attempt, at Green Bank, West Virginia, in 1960, had failed to make any such contacts—for one reason, the equipment used was inadequate—the Project Cyclops team was endeavoring to assemble an immense battery of radio antennae to use in putting a wide band of frequencies (of 200 megahertz) to exhaustive tests, sweeping through space over a range of 1,000 light years from the Earth.

One year before, in September, 1971, a similar conference had taken place at the Byurakam astrophysics observatory in the Soviet Republic of Armenia. Called by the U. S. National Academy of Sciences and the U. S. S. R. Academy of Sciences to discuss "communications with intelligent beings in outer space," the gathering included many of the world's ranking exponents of astronomy, physics, chemistry, biology and the human sciences, including two Nobel Prize winners.

Often the main purpose of such conclaves aims to offer hard-pressed scientists a brief respite from the travails of their university routine in order to discuss some new development or other, presumably urgent. The Byurakam meeting, however, was no such convivial affair. Not the first of its kind to be held in the Armenian city, the 1971 conference exerted a profound impression on the delegates present.

Back in the United States, Carl Sagan, American organizer of the event and a specialist in studies on life in outer space, and his colleague Frank Drake, who had conducted the Green Bank tests, reported to the U. S. National Academy of Sciences on the work accomplished at the meeting. In their account, the two men reviewed the probability of extra-terrestrial life in the light of the discussions they had attended.

Both Sagan and Drake agreed that the same evolutive process which developed life on Earth and eventually led to the rapid expansion of our technological civilization must have occurred under similar conditions elsewhere in the Galaxy. This conviction, upheld by many other scholars, among them the eminent British astrophysicist Fred Hoyle, rested on two factors: the vast number of solar systems in the Milky Way and the unity of physical and chemical laws which the cosmologists have discovered govern the entire Universe.

If we examine this newly-revealed galactic homeland, we find that it encompasses a huge colony of between one and two hundred billion stars. The mind-boggling prospect of such an infinite expanse of space is beyond our limited comprehension; as we contemplate the enormous vault of the Milky Way overhead, our sense of distances is so earthbound that we find it impossible to grasp such Cyclopean dimensions.

Modern astronomy, however, is beginning to acquire a fairly accurate idea of this incredible aggregation of gyrating stars. In diameter, our Galaxy extends over a spatial expanse measuring more than 100,000 light years, each light year being equivalent to approximately 6,000 billion miles. It resembles a fried egg, with the yolk, some 15,000 light years across, representing a heavy concentration of stellar matter. On the rim of this nucleus lies the solar system. Outside the galactic disk, more modest groups, consisting of stars huddled relatively close together, are caught up in the gyration of the Galaxy; hence, they, too, belong to our system. Meanwhile, two small satellite galaxies, known as the Clouds of Magellan, have been discovered at a distance of 150,000 light years away from our stellar world; these are its nearest galactic neighbors.

But the Universe stretches far beyond the Milky Way.

Indeed, our Galaxy itself belongs to a system which includes fifteen other galaxies, with a total radius of some 3 million light years. This "local group" rotates around a central point which lies between the Andromeda Galaxy and ours. The conglomerate of galaxies nearest to our "local group" is approximately 30 million light years away, while the most remote galaxy thus far detected is 6 billion light years from ours. Thus, the light rays it sends us originated something like a billion years before the creation of the Sun and the Earth.

In short, the Milky Way, with its more than 100 billion stars, is only an infinitesimal fraction of the great Universe, which harbors several billion galaxies[1], of which about 12,000 have been catalogued. Specialists in galactic studies estimate that the cosmos counts at least 10^{20} stars, or 100 billion billion suns at the minimum.

This staggering plurality of worlds is the mere beginning of a subject we can only understand through the new knowledge the world's astronomers have acquired concerning the evolution of the stars and the stellar systems. Until not very long ago, specialists believed that the different cosmic processes are brought about by peculiar happenings, but these assumptions have all given way to evolutionary theories which ascribe a powerful unity to the whole physical world.

For example, we were told that the creation of the Sun's planets was the result of a unique accident in space: some foreign celestial body had either come into violent collision with the Sun or else passed extremely close to it on an aberrant course. This theory erroneously singled out our solar system as a unique case in the annals of the heavens. Today, however, we know that giving birth to a planetary escort is normal procedure in the genesis of solar systems.

[1]The powerful telescope on Mount Palomar could photograph a billion of them, while the new Russian telescope, 20 feet in diameter, could photograph even more; but it would take a number of lifetimes for the astronomers to complete the work.

Hence, the same process that created our system must have occurred with innumerable other stars bearing the same characteristics as did the Sun early in its career.

Moreover, modern microphysics has paved the way to this new orientation. Thus, thermonuclear fusion explains the amazing stability of the solar furnace's combustion, while the scientists' wholly revised pattern of the Sun has, in turn, guided them to a new understanding of the combusion process common to all stars.

Such progressive discoveries of the laws governing the life of all stars have lent growing acceptance of the likelihood that solar worlds akin to our own must indeed exist elsewhere in the cosmos.

In the wake of such persuasive evidence, the cosmologists have abandoned the hoary dogma that the star called Sun is a prodigy in the Universe which claims a total monopoly on such phenomena as life.

Still, the fact remains that such a faith, upheld by specialists who view the problem from their stratospheric stance while the layman sees it from a sharply restricted outlook, is only a philosophical assumption rather than a scientific reality subject to concrete detection and measurement. All the same, before they could advance their hypothesis, the researchers had to have unquestionable proof that other stars besides the Sun are accompanied by planetary escorts.

Can Other Planets Nurture Life?

Of the billions of stars we know exist in space, have any of them, apart from the Sun, spawned a retinue of planets on which life could conceivably generate and develop? And if so, how can we verify the fact?

Direct visual observation is virtually impossible. Given their tremendous distances from the Earth, even those planets in orbit around the stars nearest to us are so shrouded in darkness that no telescope, however powerful, can spot them. Furthermore, any brilliance they might give off would only be confounded with the radiance of their parent suns.

The difficulty of detecting planets four to ten light years away becomes readily understandable when we realize that the two most remote planets in our own system, Neptune and Pluto, were not discovered until 1846 and 1930 respectively; yet Pluto is a mere 3.5 billion miles from the Sun and only a few light hours from the Earth.

But where direct optical observation is useless, other indirect methods can yield solid evidence that such planets do indeed exist. Through spectral analysis of the infrared radiations surrounding young stars, astronomers can detect clouds of gas and dust (the lithium lines are relatively easy to discern). During the genesis of our solar system, there were components of the solar cloud that eventually gave birth to the Sun's planets. The vast cloud formations seen rotating around young stars indicate the likelihood of planetary offshoots.

Another indirect procedure leads us to the same conclusion. Through spectral analysis the astronomers can measure the rotatory speed of stars by observing the different shifts in the two opposite poles of a stellar disk. The astronomers have established that aging stars turn slowly, while new, extremely hot stars revolve at high speeds. In seeking an explanation for the loss of momentum in adult stars, the scientists discovered that they have transmitted a part of their kinetic energy to their satellites. This is precisely what happened when the solar system came into existence. The Sun originally must have rotated on its axis at the rate of several hundred miles per second; but gradually transferring a good part of its angular momentum to its planets, it slowed down, and now rotates at the speed of only 1.2 miles per second.

The slow rotatory velocity of many adult stars favors the assumption that planetary escorts most probably exist in outer space; and adult suns account for approximately two-thirds of all the stars in our Galaxy. This reasoning, however, is by no means conclusive. Without spawning a cortege of satellites, young stars can also cede movement by expelling some of their gas by a magnetic mechanism.

Recently, still another indirect method actually led to the first discoveries of planets beyond our solar system. Astronomers succeeded in identifying them by measuring the perturbations they caused on the course of their parent stars.

This technique has been widely practiced to ferret out the invisible companions of visible stars in a multiple system. Indeed, suns are not always solitary orbs like ours; between 30 and 50 percent of those in our Galaxy exist in double or triple formations, if not more. Some of them, however, are either too small or too old, hence lacking in brilliance, to be perceptible. Astronomers, therefore, seek them out by analyzing the course of their companion suns. The two planets most distant from our solar system to be thus far discovered were signalled by these means before they were ultimately found, after astronomers had puzzled over the mysterious perturbations recorded in the course of a sister planet.

About the time François Le Perrier was calculating Pluto's position, the German astronomer Friedrich Bessel —who won acclaim when he perfected the method for measuring the distance between the Earth and a star—was perplexed by the strange anomalies in the movement of Sirius, the most brilliant star in our firmament. Bessel believed that some unknown neighboring star must have caused them. But an American astronomer cleared up the mystery a few years later when he found the intruder to be not a planet but rather a dwarf sun which, having lost its combustion, was dying. The same method resulted in the discovery of Procyon's companion sun, likewise a dwarf. At the end of their careers, such stars lose both their energy and radiance, and sometimes assume the aspect of a huge planet like Jupiter.

It required extremely long and difficult observations before the astronomers could make their first discovery of a planetary system. Of necessity, they had to settle on a sun not too far away, and keep a sharp eye on its galactic movement. All stars, of course, are caught up in their Galaxy's gyration, but their displacement will be perceptible only if the terrestrial observer focuses on those which are not too remote. After Proxima Centauri, the Barnard Star, slightly more than six light years away, is the nearest to us. Initiating his survey of this star in 1916, Professor Peter Van de Kamp, of the Sproul Observatory in the United States, could not verify the results of his observations until 1961. Ten other astronomers took part successively in the experiment, and several of them passed away before the positive outcome could be confirmed. During these years, the team took 2,413 exposures of the star, which revealed displacement anomalies of no more than 10 microns. From this evidence, they deduced the presence of a planetary body bigger than Jupiter.

After six more years of observing and calculating, Prof. Van de Kamp was obliged to modify his conclusions. The Barnard Star, he announced, was accompanied not by one but *two* planets, each with a mass comparable to Jupiter.

These discoveries by no means signify that the Barnard Star is host to these two satellites alone. There may well be others, but the observation techniques thus far developed cannot yet discern planets of such modest dimensions as the Earth.

By analyzing aberrant trajectories, the astronomers have so far succeeded in identifying eight dark bodies associated with stars less than 20 light years away from us. Most of these are "megaplanets," as big or bigger than Jupiter. (Our nearest neighbor, Proxima Centauri, a triple solar system[1] less than 4.5 light years away, apparently has no escort.) These planets are:

The single Ross 614 planet, eighty times bigger than Jupiter—and, in fact, an abortive sun—is the only one

[1]Double, triple and even sextuple stars can be accompanied by planets.

star	*distance in light years*	*size of planet by comparison with Jupiter*
Barnard's Star single star	6+n	two planets: B1: 0.8 Jupiter B2: 1.1
Lalande 21 185 single star	8.2	one planet: 30
61 Cygnus (The Swan) double star	11	one planet: 1.5
BD + 5° 1668 single star	12.4	one planet
Krüger 60 double star	13	one planet: 9
RD + 20° 2465 single star	15.4	one planet: 32
Cassiopeia double star	17.9	one planet: 10
Ross 614 single star	13.1	one planet: 80

outside our solar system to be discovered by optical observation, this through the powerful Mount Palomar telescope.

Astronomers interpret the data collected on Barnard's Star as indicating three planets and other residual bodies, the first planet being somewhat bigger than Jupiter, the other two smaller; but Van de Kamp perceives only two. For present purposes, however, the discovery of planetary retinues is more important than their particular characteristics. Thus far, no planet of the terrestrial (telluric) type has been detected, nevertheless it is possible to analyze the measurements of trajectory perturbation by several methods.

Today, the Sproul and other research teams are studying more than 300 stars within a radius of forty light years, with the help of new and more advanced techniques. The first results of their indirect detection program have shown that a planetary retinue falls within the normal process of a star's evolution. This probability has been brilliantly confirmed by theoretical demonstration, in which the process of planetary formation is reconstructed by computer simulation, beginning with the accumulations fortuitously provoked in a mathematical model of the original solar cloud. These simulations have approximated planetary systems very like our own. When simulated, the material of the original cloud obeys the same laws of apportionment that govern our solar system.

Thus, it appears that planetary systems are not the exception in our Galaxy, indeed we can reasonably assume that they number in the billions. But only stars which, like the Sun, claim a stabilized combustion can nurture life on their satellites. This, in fact, would encompass the greatest number of stars thus far catalogued, including the red dwarfs—stars older than the Sun—which are fully capable of fostering life on the planets that escort them.

Any attempt to estimate the number of planets on which living organisms can evolve would hardly be realistic. The example set by the only system we know—our own—shows that life makes very strict demands of its habitat. Nevertheless, the multiplicity of planetary systems that apparently abound in the Universe offers good grounds to suppose that it has taken root and thrives elsewhere in infinite space.

Apprenticeship of Life

Of the planets thus far detected in orbit around several of our nearest stars, we know nothing more about them than their dimensions and the characteristics of their trajectories. If we also want to know whether life can exist on them, we would be well advised to base our conjectures on a review of life's vicissitudes on Earth. Since spectral analysis has shown that the whole Universe is composed of the same elements that prevail on Earth and that the same chemistry and laws of physics obtain everywhere in space, it is reasonably safe to assume that the life-generating procedure is much the same throughout the cosmos.

Ever since that very ordinary star we call the Sun first lit up the sky some five billion years ago, its combusion has remained remarkably stable, and astronomers estimate that it will continue to transform its hydrogen into helium for still another eight billion years in the same, or very similar, conditions.

While the new Sun was expelling its thermonuclear heat, the cloud of gas and dust surrounding it began to contract and solidify, thus giving birth to its planets. As the nebular ring cooled, it created two kinds of satellites. One consisted of small, solid planets in relatively close proximity to the Sun, which threw off light elements such as hydrogen and helium that had constituted the essential matter of the original clouds, then hardened into crusts of silicate and metals. Farther away from the solar globe, huge planets took shape, retaining enormous quantities of light gases from the outer parts of the nebular mass.

At the time of its genesis, the Earth was still enveloped in its primeval atmosphere, combining water, ammonia, methane and molecular hydrogen. These first modest combinations already united the elements vital to future life—carbon, oxygen and nitrogen—supplied by the solar heat, together with an abundance of hydrogen, the greatest part of which had escaped with the helium.

This formation of the hydrosphere was a matter of great importance, since life could only emerge from water, the ideal solvent (solvents are indispensable to vital exchanges). As the fledgling planet's temperature dropped, heavy rainfalls left extensive sheets of water—the first oceans—into which the ammonia, methane and mineral salts, elements of the subsoil, dissolved. The Sun's ultraviolet rays, unimpeded by a barrier of ozone, could then initiate the chemical process that subsequently led to the genesis of life.

A celebrated experiment effected by Stanley Miller, an American astrophysicist, demonstrated that if a combination of gases reproducing the original atmosphere of the Earth is subjected to electrical bombardment, it will create amino acids and other substances that constitute living organisms. Subsequently, other scientists achieved this same reaction with ultraviolet rays and even by means of a thermal method of simulating volcanic activity. Under the action of such energies, carbon—which became the support of vital exchanges because of its adaptability in combining with other chemicals—could then begin to assemble the complex compounds that provide the basis of living matter.

The first pre-organisms were undoubtedly molecular, not cellular in type. These must have been colonies of very long carbonized chains that grew in the Earth's first oceans, forming a thick broth that fed on itself. Constant contributions from new carbonized matter, some of it probably synthesized in space, prevented this muddy broth from reverting to its primary state of minimal energy. Next, the mechanisms of duplication began to function, and the so-called left symmetry, inherent to all living structures, went into action.

In our physical world, the molecular structures develop in the three spatial dimensions, and it is possible that the same molecules adopted two different types of combinations with the same number of atoms but structurally opposed, just as the right and left hands are at once identical but different. Life, however, has a marked preference for the left symmetry (to which, in fact, all proteins belong). In this choice lies one of the mysterious frontiers that separate the animate from the inanimate. Actually, in synthetic experiments, it has been proved possible to manufacture left symmetry substances (levogyrous) by using ultraviolet rays, which figured as a critical element in the formation of the first biological compositions.

In essence, all evidence confirms the theory championed by Alexander Ivanovich Oparin and John Haldane that life originated in an evolutive chemical process, just as Darwin perceived; biologists today are within reach of explaining the inception of the self-replicating faculty, assuring the permanence of organisms. But the crucial stages in the escalation toward life remain undecipherable. We cannot account for the qualitative surge that enabled molecular structures to activate around the cell "project," the true basic organization of living matter, which Hoyle once called more complicated than a nuclear power plant. Jacques Monod thinks that this barrier will be enormously difficult to hurdle, since intermediary fossils that could enlighten us no longer exist; thus, science lacks the evidence necessary to reconstruct the missing links. The most ancient relics of organisms yet found, dating back some three billion years, already represent sophisticated forms of life.

The first organisms were anaerobes, unable to breathe oxygen, which did not yet exist, from the air. Their reactive processes were like those of yeast when it transforms sugar into alcohol. But the efficiency of this procedure was, and still is, very feeble; therefore, life was obliged to find more aggressive methods of oxydation. It had to wait until the ultraviolet rays began to create a primordial atmosphere of oxygen by dissolving a certain quantity of water. The algae, ancestors of our green plants, managed to survive without any oxydizable agent merely by expelling their own excess oxygen into the air. Thenceforth, the elements essential to all life could be continually fabricated out of such rudimentary components as water, carbonic anhydride and sunlight.

Life on Earth eventually established a two-stage cycle. First; plants stored carbon from the carbonic gas in the air by germinating a tissue of carbonized molecules with the help of water drawn from the soil, at the same time hoarding solar energy to set up an unstable system which is itself a reserve of energy. The second state: animals eat the plants and ingest their condensed energy.

The Vital Minimum

Life on Earth, therefore, emerged with the escalation of a chemistry which evolved into closed systems, endowed with an astonishing complexity of mechanisms and internal apparatuses for absorbing information, for duplication and regenerating themselves. It first appeared very early, soon after the dawn of our planetary world 4.6 billion years ago. Quite possibly it may have even begun a mere few million years after the solar system came into existence.

Life, then, lost little time in producing a primitive sample of what it would be in the far distant future.

But can it exist on other planets? And under what circumstances? Since the laws of chemistry and physics are the same everywhere, what we have been able to reconstruct of life's initial stages on our own globe will guide us in tabulating the conditions indispensable to it elsewhere in outer space.

First of all, life requires constant sources of energy. It was the Sun that assured the escalation of chemistry toward life on Earth through the energy radiated by its heat, and again the Sun that synthesized most of the carbonized matter which provided its elementary ingredients, together with the vital compounds generated by the solar heat and scattered throughout the planetary belt.

Later, when certain organisms were no longer satisfied with the energy stored by the food supplies available to them, once again it was the Sun that made it possible for them to produce new compounds of their own. Where the ultraviolet rays had ignited the "spark of life," now the light rays guaranteed permanent reserves of energy through photosynthesis. But there were also other sources of energy that played a dominant role in the development of life—the energy stored by gravitation ever since our system took form, the radio activity of rocks, both sources of heat and movement originating in the solar pivot, and galactic energy beamed to us in the form of rays. Without digressing into this complicated discourse, let it suffice to say that life emerged under the extremely generous patronage of the Sun, and on a platform itself of solar origin.

Of course, those stars, including the Sun, which provide sources of energy indispensable to life, cannot maintain a habitat for living creatures in their own right. No structure that shelters delicate metabolisms could resist surface radiations with a thermal fan ranging between thousands and tens of thousands of degrees. The so-called cold stars, with surface temperatures of fiery 3,000 degrees, must likewise be ruled out as potential habitats; but this type of star, the red dwarfs, which burn more slowly than the Sun, can also offer life a haven in their planetary systems. Barnard's Star belongs to this category.

The same question has been put concerning life in interplanetary space, where water has been discovered as well as a whole series of carbonized compounds, attesting to pre-biotic chemistry in the interstellar spaces. But space is too rarefied to foster an active chemistry; life cannot conceivably develop in the galactic void.

Life also demands a dense atmosphere and a moderate temperature, both of which will tolerate vital exchanges and sanction a stable chemical structure. Given the example of our own planet, we see that a small, dense satellite with atmosphere, water and an average temperature can encourage the inception and growth of living things. In particular, the Earth's mass had to be modest enough to permit the greater part of its light gases to escape, otherwise they would have prevented a solid crust of silicate and metals from taking form. On the other hand, it had to be sufficiently massive to retain the heavier gases—ammonia, methane and others—to nurture life.

In addition to these demands—a solid support, a continuing supply of energy at an appropriate distance from the solar furnace and a propitious atmosphere—life also requires an obliging hydrosphere with a sufficient proportion of hydrogen combined with the planet's oxygen.

Water can only exist where the atmospheric pressure is right; and here again we see the need for a minimal mass commanding a moderate temperature.

The astrophysicists maintain that the size of the mass, the position of a planet in its solar system and the chemical composition of the atmosphere are all interdependent. Such restrictive conditions imply that the size of a life-bearing planet must closely approximate the size of the Earth.

Also very strict are the thermal conditions that prepare the way for life. Organic molecules at the root of terrestrial life disassociate at 212 degrees F. Some organisms resist up to 150 degrees. In Iceland, for example, mosquitos live in geysers with a temperature of more than 130 degrees. Cold temperatures arrest the vital functions but do not destroy living structures—they only immobilize matter. In November, 1969, a streptococcus was found in a camera retrieved on the Moon by the Apollo XII crew; it had survived the icy rigors of the lunar climate for two years.

The question of cold and heat must be considered from another point of view. A fistful of sand in the Sahara will embody whole worlds of microbes, whereas the Antarctic soil, apart from its water points, is virtually aseptic. Thus, within the limits of an organism's endurance, in the bio-thermal window, cold hinders the development of life more than heat.

Did life on Earth develop by accident or a whole sequence of accidents or was the evolution of certain chemical chains in the escalation of living organisms an inevitable process? This is the first question we must confront in contemplating the possibility of extra-terrestrial life; and if the second hypothesis is the right one, then that possibility is singularly substantiated.

Considering the enormous complexity of the simplest organisms and the long series of lucky chances life had to pile up to build its structures, some experts maintain that life fought a ferocious battle for survival in a struggle it had little hope of winning. It is inconceivable, they believe, that the same miracle could happen anywhere else in the cosmos. But most biogeneticists reject this pessimistic viewpoint. Terrestrial life, they reply, held a fistful of trump cards. Numerous laboratory experiments simulating models of the primitive atmosphere and using different energy agents have all produced pre-organic molecules, such as amino acids and sugar.

But such pre-biological substances do not necessarily signify life. In their simulative processes, the experiments could only produce simple chemical systems incapable of reconstructing the catalytic and cybernetic functions that ensure the continuance and reproduction of living things. Nature had to create systems which could control their evolution.

The physicist Ilya Prigoginem, professor at the Brussels Free University and director of the Center for Statistical Mechanics and Thermodynamics at the University of Texas, has shown that open physical systems which exchange matter and energy possess thermodynamic properties that can be compared with the functions of living organisms. Both are unstable systems which achieve organization through their fluctuations. The first pre-biological molecular systems had to maintain and develop themselves by a stimulation process with the help of simple chemical procedures that encouraged the reproduction of substances of the same type, or developed reactions of control and blockage.

The biogeneticists are now in a fair way to explain how the pre-organisms managed to evolve bereft of both memory and any elaborate internal informational equipment. We begin to understand how physical imperatives commanded them to select and retain the molecules that would benefit them, and how the most advanced chemical syntheses succeeded in taking full control. Ever since Nobel Prize winner Fritz Lipmann saw how a primitive protein schema, made up of a dozen acids held together without any catalyzing agent—or informational gene—took form spontaneously, the scientists are closer to discovering what the original chemistry was that gave birth to life, a chemistry that owes nothing to chance or the miraculous.

While science will probably uncover the secret of chemistry's escalation in the not-too-distant future, crucial links in the chain are still missing. The biogeneticists are striving to reconstruct them, some by concentrating on the synthesis of pre-organic compounds from primitive atmospheres reproduced in the laboratory, others by studying chemical systems they hope will shed light on the mystery. Still others are working to find the molecules in structures less complicated than those of existing organisms, which they do by comparing living species of primitive types with others of more recent origin; it is also possible to cull valuable information from organic extracts taken from the remains of prehistoric animal life—fossilized dinosaur eggs, pre-Cambrian fossils and the like—but we must go much farther back than these zoological documents to reconstitute the chain. Some Russian researchers are examining different species to find the common trunk and reconstruct the primitive functions.

The hatching of life on Earth was an irresistible physico-chemical event; we must therefore take into serious account the probabilistic equation that indicates a wide distribution of life throughout the Universe, mindful of the fact that the Galaxy fosters so many favorable supports similar to those on Earth.

HELLO, OUT THERE

The Fabulous Men of Mars

At the beginning of the 19th century, the eminent mathematician Karl Friedrich Gauss believed that intelligent beings could inhabit the planets nearest the Earth. According to his calculations, the only possible way to communicate with these beings in outer space was by means of a mathematical language. Therefore, he proposed to convert the dense Siberian forest into a giant signboard, tracing the geometrical solution to the Pythagorean theorem. The figures would be marked in cultivations of light-tinted rye cut into the dark, sombre forest, which was to serve as a background. The sides of the triangle would be nine miles wide.

The Austrian astronomer Littrow, a pioneer in spectography, preferred the Sahara desert as his blackboard. Here he would have huge canals dug into the sand and arranged in simple geometric figures—circles, squares, triangles, etc. The canals would then be filled with water and overlaid with enough petroleum to burn brightly for six hours. Littrow assumed that such a series of flaming designs would alert the inhabitants of our neighboring planets to the fact that the Earth is populated by an intelligent species.

Today, such artless stratagems, proposed in all seriousness, strike us as absurdly naive. But the illustrious Gauss and his cohorts were captive to the limited resources of their time. Meanwhile, radio and astronautic techniques have enabled us to rake the interplanetary expanses in the vicinity of the Earth. We now know that the search for civilizations in outer space must reach beyond the solar system. Our radio antennae can intercept signals, even very faint ones, transmitted by intelligent beings from as far away as the confines of our Galaxy. But none of the Sun's planets emits intelligent signals, although Jupiter gives off powerful radio-electric impulses of organic origin.

Of our neighboring planets, only Mars could sustain living organisms or manifest chemical beginnings of life. But surveying it from fairly close range, American satellites on probing missions have found no traces of civilization on its soil. Previously, specialists had believed that they had discovered spectrum rays corresponding to organic matter, therefore only infrared detection could yield information to the exobiologists; but these observations were subsequently attributed to other causes.

Thus, the scientists have so far discovered nothing on Mars to indicate the beginnings of life or signs of any chemical process that could escalate toward life; the first satellite explorations of outer space demolished a good number of more or less reckless hypotheses on the possibility that the human race is not alone in the solar system. It might be useful here to review the negative balance sheet of their overly optimistic deductions.

At the time astronautics were introduced, our knowledge of Mars was erroneous for the most part, and distorted by all sorts of whimsical suppositions. A synthesis of our conceptions, even as late as the 1950s, will show how far afield of the facts the experts had strayed.

No doubt it is a buoyant spirit of gamesmanship that induces great patrons of science to exercise their wits in flights of fancy by way of flexing their intellectual sinews, without fretting too much about how accurate their conjectures might be. Thus, a number of *idées fixes* gave rise to all sorts of hypotheses which did nothing more than obscure the whole question of Mars. Among them was the claim, dreamed up by the Bostonian Percival Lowell, that the numerous canals which were said to streak the surface of Mars indicated a system of artificial irrigation that drew off water from the planet's poles to irrigate crops cultivated by industrious, efficient farmers. Esteemed as one of the outstanding figures of planetary astronomy, it was Lowell who discovered the existence and position of Pluto by studying the perturbations of Neptune's trajectory. The fame and confidence Lowell enjoyed lent considerable credibility to this theory when he announced it at the beginning of this century, and the mass media, then on the rise, disseminated it to a vast, worldwide public only too willing to accept it.

Today we can smile at the profound impression these "canals" made on the public consciousness when they were first "discovered" in 1864 and for decades afterward. To be sure, a number of astronomers viewed this theory with a wary eye; and in 1909, when the Meudon Observatory took telescopic photographs of Mars, the linear furrows Lowell had thought were canals turned out to be simply trails and blots of purely accidental geological origin. He had mistaken the seasonal changes of color for agricultural plantings.

But even despite this refutation, various astronomers, especially in the United States, persisted in interpreting these surface markings as canals; and the belief that living creatures inhabited and cultivated the planet continued to thrive until interplanetary explorations disproved it.

Years later, however, NASA significantly adopted maps of the canals prepared by Lowell's colleague Vesto M. Slipher to chart the zones to be photographed by Mariner IV.

It was the Mariner IV flight that put an end to all speculations of a Martian civilization. The photographs relayed home by the satellite from an altitude of 6,000 miles revealed a surface crust similar to that of the Moon, hardly hospitable terrain for a thriving civilization.

By no means were the Martians invented by Lowell and his disciples. Since the 18th century, a whole chimerical literature on Martian life had paved the way for them. The press cheerfully exploited the world's confidence in these exciting revelations and, since men were thought to exist on Mars, it naturally followed that we must make contact with them. Constantin Tsiolkovsky, a Russian precursor of astronautics, suggested lighting huge bonfires in Siberia; these conflagrations, arranged in a variety of forms, would somehow convey messages to our startled neighbors. The American astronomer Edward Charles Pickering wanted to built giant reflectors to flash signals to them. In 1901, Professor Victor Henry, of the University of Paris, rashly published a *Study of the Martian Language*.

Almost from its beginnings, radio took undisputed command of the situation. At the end of World War I, the number of radio receivers in operation suddenly multiplied, and numerous hams, including men of astute intelligence, convinced themselves that they were regularly tuning in messages from Mars.

In 1920, Guglielmo Marconi himself declared that a number of incoherent signals he was intercepting repeated the letter S—three dots in the Morse code—with puzzling frequency. But by 1924, the Martians had apparently transferred their affections to the letter H (four dots). In 1932, Karl Gunther Jansky, founder of radio astronomy, chilled this enthusiasm by demonstrating that such signals came from unidentified galactic sources in points far more remote than Mars.

Put briefly, Mars can be written off as a very dull planet.

Some years ago, the Soviet Shklovsky glowed with pride over his discovery that the two satellites in orbit around Mars were artificial; their aberrant trajectory, he announced, could only be attributed to the fact that they were hollow, therefore constructed by intelligent beings.

But photographs of the satellites taken by Mariner IX proved them to be planetary débris of wholly natural origin. Moreover, the supposedly aberrant accelerations of Phobos, larger of the two, have no foundation in fact.

In another major misconception, for a long time scientists believed that nitrogen was a massive component of the Martian atmosphere. This gas, it was thought, *had* to be the essential element and, as on Earth, it could only originate in the crust of the surface. But the survey satellites found not the merest trace of it.

We now have ample proof that most of the speculations put forth before Mars came under close, direct observation were entirely fallacious. The reason for the blunders committed in optical observation is a classical problem: the difficulty of interpretating data correctly when the detection instruments available are limited in their capacity for separation. This dilemma is the common lot of astronomers who spend much of their time attempting to decode faint, indistinct data. Today, however, the computer has become an invaluable ally in the performance of their delicate work. For example, from an input of the precise information culled by the cameras trained on the surface of Mars, the computer will refine the definition of the photographs. This is an astonishing feat whereby the information transmitted during space soundings is extended when the computer interprets the substance of the initial document.

Fitting in neatly with the convictions cherished by so many astronomers, the myth of Mars created a convenient, integral logic which seemed necessary in its time. Before the Mariner flights began, we could readily accept a convenient model of the three telluric planets, according to which Venus represented a past stage and Mars a future stage of the Earth's development—as if the evolutionary process of the planets had been determined by the varying distances between them and the Sun, and all three were following exactly the same course of development. But however alluring and comforting, this logical model was destined to vanish into very thin air in the wake of the incontrovertible scientific findings.

I have purposely sketched the history of the Martian myth to establish a firm basis of skepticism before turning to the question of civilizations in outer space.

Toward Galactic Communications

Discovering within the recent past how to exploit the whole gamut of electromagnetic waves, traditional astronomy has vastly expanded its capacity to probe the heavens. Previously, astronomers could tap only the light rays, by which, after refining and perfecting Galileo's telescope, they succeeded in charting the firmament, observing the radiations in the band of visible light. But the cosmos sends us many other *invisible* radiations—infrared, hertzian, ultraviolet, X and gamma rays—which the new science has learned to capture on the ground or above the Earth's atmosphere. These different forms of radio astronomy have gone far beyond the limits of optical observation; electronic ears and various types of detectors installed on the ground, on balloons, rockets and satellites yield considerably more information than did optical telescopes.

Nevertheless, all methods of detection on the various bands of spectral radiation are complementary, and optical observations still offer invaluable advantages. In addition to the specific information it furnishes, visible light enables astronomers to localize radio sources while radio antennae, much less accurate in establishing directions, hear well but localize badly.

The science of astronautics has improved the emission techniques to such a degree that they can now penetrate into incredible distances. Radar, for one, has scoured the soil of Venus. Today it is possible to direct emissions to a planet not only by hertzian waves but also by light rays through means of a laser, which can train an illuminated beam three or four miles wide on Mars when it comes within 35 million miles of the Earth.

The properties of radio waves are superior for galactic communications. The stars radiate the greater part of their energy optically, much less in radio waves; thus, the listener could more easily capture on that band any signals outer-spacemen might transmit.

In 1959, Fred Hoyle and his Cornell University colleague, Giuseppe Cocconi, proposed for the first time ever to marshal these modern transmission techniques for an attempt to make contact with civilizations that might exist somewhere in the Galaxy beyond our solar system.

But whether listening or transmitting, the radio astronomer can succeed only if he is fortunate enough to tune in some outer space correspondent who has mastered these same methods of communication. Indeed, when astronomers contemplate the search for contacts with extra-terrestrials, what they have in mind are civilizations more advanced than ours. The optical signals Gauss and Littrow meant to hew out of the Siberian forest and dig into the sands of the Sahara were intended for a hypothetical Martian society lacking a telecommunications technology.

Writing to Sir Bernard Lovell, director of the Jodrell Bank Observatory in England, in 1959, Cocconi avowed his belief that life is not a phenomenon reserved exclusively for our globe. It is very probable, if not certain, he said, that societies similar to our own are inhabiting other planetary systems, and that "there is every good chance that animals far more developed than man exist on some of these planets." Considering the rapidly accelerating rhythm of technical progress, it is quite possible that a civilization only a few years older than ours could have already equipped its "animals" with technological trappings more sophisticated than any we now possess. It is altogether justifiable, then, to assume that such societies in outer space are endeavoring to enter into contact with other worlds. To intercept them, Cocconi proposed using a very wide-range radio antenna like the huge paraboloid mobile unit, 250 feet in diameter, built at Jodrell Bank under the supervision of Sir Bernard (who, apparently, did not take the hint).

Hoyle, too, was certain that intelligent beings in outer space must be striving to establish communications with other worlds. To him, it was a matter of no little urgency. Man's efforts to collaborate with them should not be considered a hobby scientists liked to pursue, as it is in some quarters. On the contrary, such efforts must come within the normal perspectives of our global civilization's growing evolution.

Indeed, scientific growth and the progress of societies depend upon improving communications and an increasing dissemination of shared knowledge. A man who isolates himself within a small social, family or tribal cell and thus shuts himself off from the outside world will absorb only very limited information, hence thwart his intellectual development. The dynamic attributes of a society, Hoyle declared, can only be measured by the extent of its communications facilities. To be civilized is to be plugged in, to take active participation in common knowledge, to develop the antennae of the brain.

Consequently, a society evolving on the astral scale cannot avoid the search for communications with other worlds; this is one of the imperatives of its intellectual expansion. We earthlings have now arrived at that stage. Extra-terrestrial communities, which must certainly exist according to all statistical probabilities, are undoubtedly moving in this same direction; and surely, those which have attained our level of civilization must have turned their sights on outer space to find means of communicating with other intelligent beings in the galaxy around them.

Project Ozma

In the same year of 1959, Cocconi and another

astronomer, P. Morrison, of the Department of Physics at the Massachusetts Institute of Technology, undertook science's first study on the possibilities of intragalactic communications. Their elaborately detailed hypothesis was naturally limited by the conceits and methods of the times. To acquaint the scientific world with their project, Cocconi subsequently published a précis of their conclusions in the magazine *Nature*, one of the top clearing houses for new ideas in the domain of the physical sciences.

In their purely conjectural approach, Cocconi and Morrison named the radio antennae that would be most suitable for an experiment in galactic listening. The National Observatory of Radio Astronomy at Green Bank, W. Va., boasted an antenna eighty-five feet in diameter, which could intercept signals over a radius of twelve light years—assuming, of course, that any prospective correspondent within that range commanded radio equipment of equal potency. Only six stars, however, lay within the scope of the Green Bank antenna, a factor that promised only meager results. Another antenna, 135 feet in diameter, then under construction at Green Bank, would eventually be capable of penetrating a distance of fifteen light years and reaching some thirty stars.

England's Jodrell Bank antenna, 250 feet in diameter, could scan a distance of twenty-five light years, including 200 stars. Meanwhile, a huge stationary antenna (1,000 feet), to be built in a crater at Arecibo, Puerto Rico, and scheduled to begin operating in 1963, would range over 100 light years and some 10,000 stars.

The Cocconi-Morrison study, however, chose to concentrate on the Green Bank antenna. Of the six stars nearest to us, two of them held out high promise: Tau Ceti, 11.9 light years away from the Earth, and Epsilon Eridani (10.8). Both stars are spectral in type, similar to the Sun, and like it, about five billion years old. Their revolutions are slow, indicating that they might well be attended by an escort of satellites; therefore, one could suppose that they foster life on their planets—if they had any.

Cocconi imagined that a civilization or civilizations as advanced as ours, if not more so, could be living somewhere in their planetary system; and if such societies were sufficiently progressive to engage in interstellar communications, an antenna like that at Green Bank would have a fair chance of intercepting their signals—but only on condition that the Tau Ceti and Epsilon Eridani radio astronomers had begun transmitting in our direction at least eleven or twelve years previously. It would take that long for radio electric waves, traveling 180,000 miles per second, to arrive on Earth.

Obviously, even under ordinary circumstances, the successful outcome of such an experiment must encounter discouraging difficulties. All the same, Cocconi's essay in *Nature* fired two American astronomers, Frank Drake and William Waltman, with a passionate ambition. Both, by coincidence, were members of the Green Bank staff. Using their powers of persuasion, they obtained permission from the director of the installation, Otto Struve, to reserve the paraboloid antenna for a listening experiment requiring several months' time. Considering that all great radio telescopes are so overburdened with commitments that even high priority experiments must patiently await their turn, Struve's acquiescence demonstrated his laudable good will.

To this momentous initial attempt, Drake assigned a code name which evoked visions of that fabulous, inaccessible Land of Oz: Project Ozma.

In practice, Project Ozma posed several thorny problems. What wave length should be used? Cocconi and Morrison had indeed specified the tuning-in limits of such an experiment. A wide band wouldn't work, and if Drake and Waltman chose a wave length of more than 300 meters, the interstellar medium would absorb any signal emitted on these frequencies. Moreover, the atmospheric blanket, being impervious, would allow no waves more than fifteen meters in length to penetrate it. The minimum would be three centimeters; below that limit, the antenna would be at the mercy of atmospheric molecules, which absorb the waves.

Drake selected the wave length recommended by Cocconi and Morrison corresponding to the natural ray emitted by hydrogen. What led him to make this choice was the fact that hydrogen is the most prevalent element in the Universe, composing more than ninety percent of its matter. This cold hydrogen emits energy in the form of radio electric waves on a length of three centimeters, with each atom a sort of microscopic radio sender, regulated to emit on one or two particular wave lengths under given conditions.

Certainly, any advanced society could not have failed to detect this primordial element; and since their radio astronomers presumably had mastered the telecommunications techniques, they would know its characteristic wave length.

Today most radio astronomers concerned with outer space communications agree that the radio electric signal of hydrogen probably constitutes a common denominator among extra-terrestrial civilizations; therefore, the twenty-one-centimeter wave length should be a sort of universal standard. Yet Drake did not hold to this exact wave length. Instead, he tested the whole span of a 400-kilohertz band, equivalent to ten transmission bands of a standard radio transmitting station.

Project Ozma lasted for three months, from May through July of 1960, but without luck. As the final bulletin ruefully announced, "No signal of outer space origin was discovered during these preliminary observations."

Project Ozma aroused a storm of protests, mainly against Drake's choice of wave length. Critics declared that any incoming communications on the twenty-one-centimeter band would have been jammed by the static buzz of interstellar hydrogen concentrated on the general rotation of our Galaxy.

But inasmuch as Drake had focused his efforts on two

relatively close stars, this complaint was by no means valid. All the same, the twenty-one-centimeter wave length is not calculated to give the best listening results in any attempt to hook up with more distant stars; Hoyle suggested a higher frequency. Meanwhile, most radio astronomers who have grappled with the problem believe that this natural standard must necessarily be adopted for galactic contacts. They are convinced that outer space astronomers testing radio electric waves for the purpose of communicating with other worlds could not have failed to experiment with it.

Other experts have proposed a solution that would reconcile listening needs with the requirements of a common standard: radio astronomers on Earth and in outer space should find a suitable multiple of this wave length, since signals transmitted on such a universally accepted standard would not risk interference from galactic noises.

When Project Ozma failed to measure up to galactic reality, Shklovsky was prompt to give his explanation for its defects. "Here we have only an embryonic listening table of interstellar conversations," he commented. What is needed, he thought, was a detection system operating on an altogether different scale. But as Sir Bernard Lovell exclaimed in admiration, "It was an experiment few of us expected to succeed, but one we should all have liked to try."

INTELLIGENT WORLDS TO DISCOVER

The Life and Death of Galactic Societies

Project Ozma was followed by similar experiments in Great Britain and the Soviet Union, in the latter country under the direction of Kardashev and Patchenko, both members of the Sternberg National Institute in Moscow. Despite the negative results of these early efforts, astro-scientists in different parts of the world undertook to work out comprehensive hypotheses on the probable existence of advanced societies in the Galaxy, including their number, the distances separating them and their longevity. In the United States, these studies were coordinated during a parley at Green Bank in 1961, when a group of astrophysicists, radio astronomers and exobiologists met to conduct a post-mortem on Project Ozma to find out why it failed. During the proceedings, they devised an equation to provide what they considered the parameters for calculating the number of galactic populations equipped for spatial communications. The equation:

$$N = Te + Fp + Ne + Fv + Fi + Fc + Dc$$

N represents the number of advanced societies in the Galaxy;

Te: the average annual rate at which new stars take form in the Galaxy;

Fp: the fraction of stars with a planetary system;

Ne: the number of planets per system with an ecology favorable to fostering life;
Fv: the fraction of these planets on which life has actually emerged;
Fi: the fraction of life-sponsoring planets on which intelligent beings or life in some form has developed;
Fc: the fraction of planets harboring intelligent life which has matured into advanced civilizations;
Dc: the average life-span of these civilizations.

As Carl Sagan pointed out, the only quantity we can estimate with a certain degree of accuracy is the first; the average annual birth rate of new stars. This figure is calculated by dividing the number of stars (over 100 billion) by the estimated age of the Galaxy—more than 10 billion years—or a yearly birth rate of 10 stars.

The probability rate of the other quantities in the equation must depend upon hazardous extrapolations of the Earth model. The scientists, however, are beginning to assemble a mass of objective data on the planetary systems, and the planetologists generally agree with Sagan that there is an infinity of planets ecologically favorable to life, on an average of one to a system.

On Earth, life evolved through intelligence, which enabled man to adapt himself to many harsh rigors and overcome them; indeed, in his struggle for survival, his intelligence compensated for certain physical shortcomings inherent in the human species, and ultimately assured him control over the planet. From this case in point, then, it is reasonable to accept the postulate that intelligence is a natural, indispensable factor in the normal process of evolution; and if it exists on this globe, then why not elsewhere in the Universe, like so many other cosmic elements.

Assuming that it can and does, if we want to enter into contact with other civilizations in outer space, the first problem to be solved is their longevity. When a society's duration is much shorter than the life of its sun, then our chances of communicating with it will be sharply reduced. In measuring the life-span of civilization, using a method based entirely on probabilistic analysis, Von Hoerner concluded that the average duration is no more than 6,500 years. His assessment may be pessimistic, nevertheless his system need not be discarded if other fundamental parameters are applied to it. For one thing, Von Hoerner did not take into account the benefits that cybernetics, informatics and astronautics could lend to an advanced civilization in its struggle for survival, unlike the Soviets, who put much store in them. Although these factors certainly smack of science fiction, we can now perceive their vast potentialities.

Why do civilizations die out? There are several reasons. For one, a civilization, like a human foetus, can be aborted at birth, an eventuality to which Von Hoerner refers. But if its birth is normal and its intelligence is properly "nourished," then its intellectual development will be irresistible and it must eventually progress to the state of our technological civilization, if not surpass it.

In the West, the notion is widespread that civilizations are doomed once they master thermonuclear technology. Certainly, if intelligence can forestall a society's decline and fall, there is still the threat that some nuclear mishap or other may befall it; and this is only one of the risks any technological-industrial civilization must run. There are also bacteriological, chemical, genetic and other mortal dangers it can face. On the other hand, many believe that technology and science can find remedies to the ills they engender and avert these perils of self-destruction.

Likewise, physical and mental degeneration can send a society on the way to its doom. Or it will come to the same untoward fate if it abandons its superior technologies. Among other agents of global mass suicide, Shklovsky believes, is a surfeit of information. Certainly, the development of intelligence and the rise of technological civilizations are only possible when shared knowledge is widely propagated; Hoyle considers this the mainspring of man's progress. But the mounting volume of facts and figures dinned into the collective mind of an advanced society can become so overpowering that its members will be unable to withstand the onslaught. The more excessive information people absorb, the more the glut will thwart their normal processes of rational decision and action.

To be sure, computers will take over and relieve the individual of the burdensome need to master vast stores of information—but at a crippling cost. It is not comforting to imagine the behavior of human beings no longer conditioned and allied by a fund of common knowledge, which plays an essential role in conditioning their mental structure, with the mind functioning only as a clearinghouse for transient memories. Yet there are disturbing indications that earthlings are already inclining toward such a kind of destructurization, which could eventually trap them into a slothful state of indifference and inactivity. Intelligence was originally programmed dynamically to ensure the individual's survival, but only if he stores a reasonable accumulation of mental data necessary to his adaptation.

In his attempt to estimate the possible longevity of galactic civilizations, Von Hoerner listed five eventualities: the destruction of all life on a given planet; the destruction only of its highly evolved inhabitants; a mass degeneralization of the population's physical and intellectual faculties; a growing apathy toward science and technology ending in indifference; or continuing progress and a long duration.

According to Von Hoerner's estimations, however, no civilization can endure indefinitely. He allows it five chances out of one hundred to last one hundred years; sixty out of one hundred to last thirty years and to collapse with the downfall of its superior leaders; fifteen chances out of one hundred that it will degenerate in thirty thousand years; and twenty chances out of one hundred that it will lose all interest in science and technology within one thousand years.

A true Marxist, Shklovsky took violent exception to this manifestation of Western pessimism, which he denounced as apocalyptic, capitalist negativism. But a number of scientists, influenced by the gloomy predictions of Oswald Spengler and other prophets of doom, take their cue from the model of human nature, as did the ethnologist Jacques Soustelle, and believe that supercivilizations are all inevitably marked for a rapid decline.

The Soviet scientists, more optimistic, believe that intelligent societies can exist somewhere between 100,000 and one million years.

These factors hold a primary importance in the search for communications with extra-terrestrials since our chances of making contact with them depend upon the life-span of their civilizations no less than on their number. If Von Hoerner is correct in his theory that intelligent life exists in only one solar system out of every three million, then societies averaging some 6,500 years must be separated from each other by several thousand light years. At such distances the odds are unlikely that they could exchange signals.

The Soviets, on the other hand, maintain that outer space societies are separated on the average by only one hundred light years; therefore, if they command the technological means earthlings now possess, communications among them are altogether feasible.

In a somewhat more cheerful mood, however, Von Hoerner acknowledges that most contemporary galactic civilizations are probably older and decidedly more progressive than ours; all the same, their tremendous distances from the Earth and the brevity of their life-spans would barely give us enough time to exchange a few messages, inasmuch as a single exchange could require some two thousand years.

We can only hope that his calculations fall wide of the mark because if the take-off of intelligent societies into the technological age is sufficiently rapid, by establishing a network of communications they could help each other to speed up the universal pace of development through mutual exchanges of information.

In an interesting theory concerning extinct civilizations, Shklovsky thinks that one day our Galaxy will quite possibly prove to be a rich exploration site for a future space archaeology, where generations to come will find wreckage—perhaps even including data banks—left behind by vanished populations of the cosmos, yielding information exploitable long after their demise. It may well be, he says, that space is strewn with such relics. With their hoards of invaluable information, such spatial finds would promote the sharing of knowledge on the galactic scale, and promote the "reactive effect" whereby the evolution of advanced societies would be rapidly accelerated.

Here again we move into the realm of science fiction, imagining a great communications hook-up linking the intelligent societies of the Milky Way, as Ivan Efremov foresees. But some scientists engaged in the study of the infinitely big heartily endorse science fiction as a wholesome, powerful incentive to magnify the imagination of researchers. In fact, in their spare time, a few of them find pleasant relaxation in scribbling science fiction stories—like Hoyle, whose futurist yarns are widely read and appreciated.

Expanding Civilizations' Need for Energy

If we want to imagine the evolution of other worlds in outer space, we must refer to the Earth model. Man, to maintain himself and ensure his progress, must depend upon those forms of energy available to him on his planet. With his intelligence he has discovered and put to profitable use a variety of sources, beginning with fire, which gave him thermal energy in such quantities as he needed, then animal traction, horses, camels, llamas, etc. for transportation. Later, man harnessed water power, windmills, steam, kinetic and chemical energy to produce electricity, and atomic energy. But while hunting, farming and other simple pursuits necessary to life required only human effort and muscles in the past, today our technological-industrial civilization is founded on huge consumptions of fuel energy. Energy is the barometer of development.

Where the Earth's ecological balance formerly lay within the harmless limits of crop-growing, today man is drawing upon the biosphere to such an extent that he is threatening it with irremediable destruction. Modern systems of intensive agriculture and industrial stock breeding, for example, represent a disastrous liability in energy consumption. The American methods of food production burn up so much power that 27,000 calories are required in chemical farming and mechanized processing to yield a ration of only 3,000 calories. The waste is staggering.

Yet our industries are burning up increasing quantities of power to turn out more and more synthetic goods. With the needs of the masses rising vertiginously in an unarrestable process, we cannot avoid the perils of a steep escalation. The demand for energy, particularly in the area of food production, must climb sharply to keep pace.

If mankind continues to pursue this course, the coming crisis will inevitably lead the population of the Earth to a major disaster, according to John A. Carver, Jr., Vice-President of the Federal Reserve Commission in Washington. Unless we pull in the reins, he warns, before the end of this century our supplies of energy will be so exhausted that economic development throughout the world will come to a halt. The furnaces of Pittsburgh's great steel mills will go cold, the assembly lines in Detroit's motorcar plants will stop moving; in Los Angeles, those robust individuals who survive a hypothetical epidemic will be driven by desperation to eke out a bare subsistence by cultivating their lawns, their neighborhood parks and the strips of soil flanking their highways. In London, the miles of docks along the Thames will be deserted, while offices, factories and homes everywhere will be plunged into darkness.

Carver painted this grim picture in 1972 to arouse the public, which had not yet understood that the resources of the globe are limited. But even if the world's population were to revert to simpler, more economical and less wasteful practices, new sources of energy must be found for our civilization to manage to survive, to continue progressing and to rebuild its ravaged biosphere.

Potential resources do indeed exist. Apart from nuclear fission, which leaves corrosive pollution in its wake, one valuable solution to the problem is thermonuclear fusion, producing power harnessed by miniature, enclosed suns, built on Earth by man. But we must also turn to our natural Sun and its huge potentials, which are only minimally exploited today. By way of illustration, the photosynthetic process consumes only forty billion kilowatts, whereas the Sun sends off 173,000 billion kilowatts. Today we are producing and consuming ten billion kilowatts, with ninety-nine percent of this energy coming to us from the Sun; unless there are hidden properties in the atom still to be discovered, this bounty should satisfy us for some time to come. But then, calculating even a very small demographic growth rate, this Earthbound solar energy will be depleted within the next several centuries, obliging us to import part of our energy requirements from space and from other planets.

The science of astronautics has shown us how this can be done. Already some space engines have been equipped with photo cells that capture solar energy directly; and now under examination is a system of adapting the same technique to provisioning the world of the future. NASA has been studying a projected space power station which would transmit resources of electric power to Earth via microwave channels in sufficient quantities to supply the entire needs of New York City.

In the centuries to come, however, as Tsiolkovsky

predicts, the swollen population of the Earth and its gargantuan demands will oblige our descendants to scour the Martian and lunar platforms for new materials and eventually to colonize space in the relative proximity of our planet. This prospect is by no means a fanciful vision. For now, however, there is no economic motive to warrant such futuristic enterprises; ample resources still remain to be found on our planet, and investors will hit on less costly techniques to unearth and exploit them. Meanwhile, the science of astronautics is beginning to uncover the clues that will pave the way to man's future expansion into the solar system. Spatial transportation, once the prohibitive costs have been brought down to a viable level, will introduce new achievements in cosmonautic exploitation. Surely the time will come when, with the Earth's treasures depleted, man will learn how to annex planetary space to the biosphere.

To the Soviets, this future course is inevitable; and if the population of planet Earth eventually finds means of branching out into space, then why wouldn't other more advanced galactic communities have done the same? No doubt they have, said Kardashev back in 1964, when he introduced the concept of interplanetary and interstellar colonization, dictated by the need for energy which supercivilizations could not satisfy on their home grounds.

Communications and Energy

Kardashev classified energy-famished societies in three types:

A civilization of type I, at something like our stage of progess, occupies the entire surface of its planet and relies entirely on the energy sources of its territory.

A civilization of type II exploits the total resources of its planetary system, colonizes other planets and derives the utmost benefit from its solar heat. By Kardashev's deductions, a technological community's conversion from type I to type II must be fairly rapid: several thousand years will do it. The Earth's population now stands at the threshold of its transmutation to type II.

A supercivilization of type III has already expanded beyond its solar system. When it has occupied the whole of its planetary system, it will look for the *Lebensraum* it needs in the Galaxy beyond, where it can utilize the sources of energy provided by tens of billions of stars.

This conversion to the galactic scale will inevitably occur at a snail's pace: the vast dimensions of extraplanetary space necessarily impose a slow rate of change. Thus, a supercivilization setting out on such an adventure will find itself in a position much like that of hardy pioneers marching forth on foot to colonize the wide open spaces of the Far West or Siberia. The passage from type II to type III, then, could take tens of millions of years. Altogether, for a planetary society in our stage of development to achieve galactic status could require several hundred million years.

In the light of these considerations, the Soviet researchers have been trying to figure out what telecommunications methods societies in varying phases of progress might have adopted to establish space dialogues with other societies, and what our chances of making contact with at least one of them might be.

Still commanding only limited sources of energy, type I civilizations will be constrained to transmit their signals on very narrow bands and train their radio antennae only on their nearest stars. Our chances of capturing their emissions are extremely small. The technique Drake used in Project Ozma proves to fall far short of galactic reality.

Our chances, however, become much brighter if we aim for contact with some supercivilization. With far more potent energies at their command, its radio astronomers can transmit into all directions on the whole band of radio-electric waves.

According to Kardashev, supergalactic signals resemble those of a very narrow radio source because their emissions originate in a far-distant platform, which will appear to us like a pin-point; on the other hand, its spectrum will be very close to the noise coming from a natural source. Meanwhile, an intelligent transmission on a very wide spectrum would be more intense in the region of the decimetrical waves, the most favorable for listening, and weaker on both sides of the frequencies on which the noise increases. Such an artificial transmission can easily be distinguished from the spectrum of the natural radio sources, whose power generally wanes with the wave length.

Kardashev also declared that the intensity of an artificial transmission would of necessity vary with time. This will become evident as soon as the operator tries to attract attention or transmit information.

His enthusiasm for his theories trapped him into making a curious blunder. In 1966, another Sternberg astronomer named Cholomitsky was studying radio source CTA 102, previously discovered by Americans, when he noted that it was emitting signals with an intensity that varied with a puzzling regularity. Hitherto, no such phenomenon had ever been known to radio astronomy. The emissions bore a striking resemblance to the very same "intelligent" signals Kardashev had been discussing.

The Sternberg astronomers promptly called a press conference, at which Kardashev declared that the signals could be coming from some artificial source. Shklovsky prudently took a noncommittal stand. Professor Martinov, director of the Institute, announced that his staff would investigate the matter and verify the periodic emissions.

Several days later, the Institute made known its findings. CTA 102 was a quasar some ten billion light years away, or at the very extremity of the known Universe. In view of the enormous distance, the emission must have been so potent that it could not be attributed to any outer space transmitter.

Thus, Kardashev's hopes of making the first contact with intelligent beings in some other world were sadly dashed. But some astronomy writers, eager to arouse public interest in such matters for their own benefit, still cite CTA 102 as valid proof that we are already in communication with extra-terrestrials, although they carefully refrain from quoting the conclusion arrived at by the Sternberg scientists.

Later, the staff of the Cambridge observatory in England encountered a similar situation, but one that ended in a wholly unexpected denouement. Charged by Professor Anthony Hewish, director of the Cavendish Laboratory, to analyze tapes recorded from the observatory's radiotelescope, a young student, Jocelyn Bell, was puzzled by certain periodic impulses she heard, and sought to locate their origin. She was convinced that they had nothing to do with terrestrial parasites—radiophonic perturbations or radio noises on Earth—but rather with distant signals from somewhere in the Galaxy. Alerted, the laboratory's astronomers promptly began to make the necessary verifications. They noted that some mysterious source was sending out impulses at several hundredths of a second, similar to those of an atomic clock. Was it a message from another planet? Sobered by the Sternberg gaff a decade before, Professor Hewish preferred to keep his investigation under wraps until he could identify the phenomenon. Meanwhile, periodic signals of the same kind were picked up from other regions in space. Several months of investigation led to the discovery of pulsars, which send out intermittent signals like a rotating lighthouse beam, possibly originating in the vibrations of stars in their death throes. For this achievement, Professor Hewish was awarded the Nobel Prize.

His staff had good reason, of course, to rejoice over this victory, but perhaps not without a tinge of disappointment; for one exhilarating moment, they had been so confident of winning the race to intercept the first messages from outer space that they even tagged their nonexistent interlocutors with the nickname LGM 1, for Little Green Men 1.

A Galactic Language

Before we can enter into communications with intelligent races on other planets, we must settle on a language our correspondents will be able to understand, just as they must communicate with us through some medium we can decipher. Among the first to cogitate this problem was Gauss, Littrow and others who, a century ago, proposed the language of mathematics. Since the logic of numbers and the principles of physics are now known to apply not only to the planet Earth but to the whole cosmos, today most scientists concerned with the question endorse this recommendation. Mathematics, which can be deciphered and translated from symbols, should be, by all logic, the natural lexicon of space dialogues.

It must be a logistic language that holds the key to its own decipherment. Some mathematicians believe that such a common tongue must be constructed progressively, beginning with the simplest statements of fact: one plus one makes two, this is right, that is wrong, and so on. Proceeding from this point of departure, step by step a whole vocabulary would emerge in symbols our correspondents would come to identify, mastering it as they acquire the fundamentals of our terrestrial knowledge.

Hans Freudenthal, a mathematician at the University of Utrecht, has attempted to forge such a cosmic esperanto which would enable galactic races, which have nothing in common except reason, to communicate. But what he calls his *Lincos* has raised objections from others on the grounds that each new idea introduced must be explained by a series of references to symbols previously transmitted, which our correspondents may or may not have grasped correctly. Moreover, some specialists doubt the feasibility of imposing a closed system on them—for surely, we are closed in a world of mental representations that belong exclusively to us. To Soviet radio astronomer B. I. Panovkin, communications are possible only among communities which have had similar historical experiences as a basis for mutual reference.

Obviously, to comprehend a totally alien, isolated system, the reader must have the key to it. Champollion was unable to decipher Egyptian hieroglyphics until the discovery of the Rosetta Stone, on which one message was engraved in three languages. It is not inconceivable that the logic of mathematics can eventually supply such a key to its symbols and therefore make intergalactic communications possible.

Another system might be the transmission of images, formulating an analytical language of conventional signs. In a demonstration of this method at Green Bank, Frank Drake asked his listeners to imagine that they had picked up a series of incoherent signals from some unknown point—brief signals separated by more or less lengthy intervals. A close scrutiny of the tape-recorded message would show that the same sequence is repeated insistently, giving clear evidence that the origin of the signal is not accidental but artificial. Studying the basic sequence, the examiners discover that the duration of the intervals coincides with a multiple of the sound signals' duration. They can therefore conclude that the signals constitute a message ciphered in the binary code, using two symbols—the simplest of all codes—expressed in the figures 0 and 1, with each interval lasting as long as the 0 symbol. The 0, of course, indicates a blank space, the numeral 1 a black space.

This will form a patter like 1000110010, or something similar.

Drake then told the members of his audience to decipher the message, which a good number of them succeeded in doing without much difficulty. This is how they reasoned:

The complete sequence converted into 0 and 1 signals totalled 1,271 signals. Since astronomers are all mathematics experts, they soon saw that the number 1,271 is the multiple of 41 times 31. These two primary numbers are obviously an instruction to arrange the signals in 31 lines

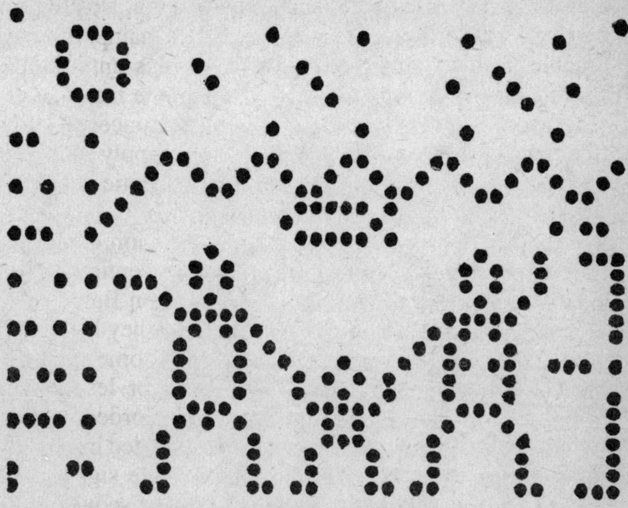

2(11), 3(101), 4(111), 5(1001), 6(1101), 7(1011), 8(1111), 9(10001).

of 41 elements each, similar to the division of lines on a television screen. This, in fact, was the same system followed when interplanetary projectiles photographed the surface of Mars; they transmitted their exposure point by point, and the complete images were reassembled on Earth.

The whole sequence invests Drake's hypothetical image with "rich information," as Shklovsky put it, which the astronomers could make out with no trouble. It delineates three humanoids, not unlike earthlings, posed in a family group, the basic social cell.

In the upper left corner of the diagram, a circle simulates a sun capping a column of eight dots, evidently its cohort of planets. To the left of each dot is an example of binary notation, giving the order of the planets: 2(11), 3(101), 4(111), 5(1001), 6(1101), 7(1011), 8(1111), 9(10001).

This gave the number labels adopted by Drake's imaginary extra-terrestrial correspondents.

The male is pointing to the fourth planet, the one he inhabits. Opposite the third dot, an undulating line would appear to signify that water has been found on that neighboring planet, while the design below it resembles a fish. Since he knows these facts about the third planet, obviously these humanoids can travel from one planet to another. They are living in their age of astronautics.

The dots along the top of the design and to the right of the sun spell out his planet's chemistry. Recognizable from left to right are the features of hydrogen (a nucleus and an electron); carbon, with its cortege of six electrons, four of them peripheral; and oxygen, with its eight electrons. Thus, his world, so tersely summarized in the diagram, is very like our own.

In addition, the chart enlightens us on the stature of the humanoids depicted. The measuring apparatus sketched at the extreme right suggests the number 11, again in the binary code (11101). Since the transmission was intercepted on 21 centimeters—the wave length of hydrogen, which could well be a universal standard, given its prevalence in the cosmos—we can estimate that the two adult humanoids stand 2.31 meters tall—in other words, giants.

At the end of the female's arm is the figure 6(1011), which could mean that she has six fingers instead of five. These extra-terrestrials apparently do their counting on the duodecimal rather than the decimal, system.

Communications via images would require a gradual procedure in multiple stages. In the primary stage, the sender would draw the attention of interceptors by a simple call, a signal that would be distinguished from natural emissions by its uncommon structure; then he would begin to convey information in elementary terms.

While Drake's concocted message may seem like an amusing rebus to entertain his associates, it is a good example to show how a complicated store of information can be conveyed in very terse phraseology.

Yours Truly, Mankind

With the launching of Pioneer X on March 3, 1972, NASA took advantage of the occasion to post man's first missive to dwellers in other worlds. The message is engraved on an anodyzed aluminum plaque measuring 8 1/2 x 6 inches attached to a point on the vehicle's flank where it is wholly visible, yet relatively well protected from interstellar dust.

The idea of conveying mankind's good will tidings to other planets by this means occurred to one Eric Burgess, a journalist for the *Christian Science Monitor*, who was somewhat indignant that no one at NASA had thought of it. He passed the suggestion on to Carl Sagan who, in turn, proposed it to NASA. Given official approval three months before the rocket's scheduled launching, Sagan enlisted the collaboration of Frank Drake in drawing up the text. Since language would naturally be incomprehensible to any eventual addressees, the two men agreed to frame their communication in imagery, or in the form of a design giving vital information on our whereabouts in the Universe and a hint of what we humans look like. Mrs. Linda Sagan, a painter by profession and wife of Carl Sagan, was commissioned to execute the work.

Sagan wanted the plaque to carry a "message from all humanity," consequently there would be no national flag depicted, and any ulterior political motives were to be strictly proscribed. Moreover, the nude couple portrayed in the design were to bear no consistent racial features; Mrs. Sagan gave the female Oriental eyes and the man a negroid aspect, investing both with a comely appearance and an air of disarming innocence. The man has raised his right hand in a gesture of greeting and good will; we can only hope that the significance will not be taken amiss.

Behind the couple is a sketch of the spacecraft itself, meant to serve as a measuring rod to indicate the dimension of these earthlings. At the base, ten circles represent the Sun and its nine planets. The route traversed by

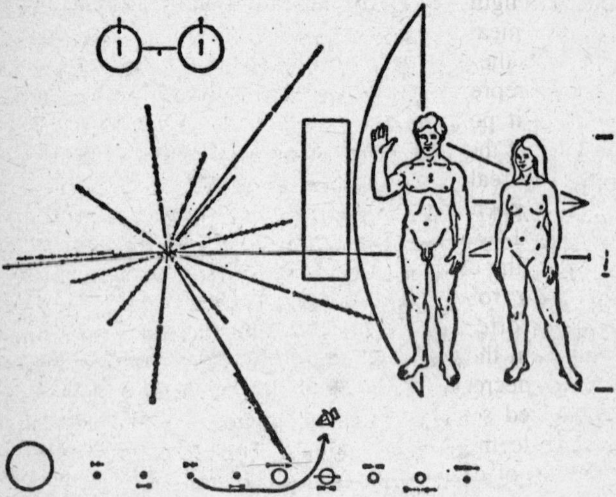

Pioneer X from its starting point on Earth is traced by a line passing between Jupiter and Saturn, then curving upwards ending in an arrow pointing to the Galaxy.

To indicate a comprehensible representation of our solar system's position among the billions of stars in the heavens, as landmarks Sagan and Drake chose 14 pulsars in the Galaxy. These radio sources emit purely characteristic signals, rhythmed by very fine impulses which cannot be mistaken for any other natural emission. In her design, Mrs. Sagan enciphered this information in the binary code by tracing the pulsars in 14 lines radiating from a central point. Her notations specify the frequency on which each pulsar was emitting at the time the spacecraft was launched. The point of convergence marks the position of our solar system.

The standard of wave length is indicated by the hydrogen atom—the most prevalent element in the Universe—which figures symbolically at the top of the design in two states of energy. Passing from one to the other is a photon; this emission, intercepted on the twenty-one-centimeter wave length, represents the basic unit of a universal system of measurement.

With the aid of the binary symbol at the right of the design representing the numeral 8, the hydrogen standard makes it possible to determine, for example, the exact height of the woman. Hence, our very first ambassadress to other realms measures 1.68 meters, or 8 x 21. Those who happen upon Pioneer X will be able to confirm this standard by checking it against the sketch of the rocket behind the couple, drawn to their scale.

Apart from fixing the Sun's position, the symbol of the pulsars offers other vital statistics. The frequency of the radiation they send out diminishes with regularity. This rate of decrease, which should surely be understood by an advanced society, will enable those who examine it to calculate the moment in time corresponding to the frequency of each pulsar. Even after a million years, any extraterrestrials who retrieve Pioneer X will presumably know how to estimate the moment in galactic history when the diagram was made because the shift in frequencies acts as a sort of cosmic clock.

The data engraved on the plaque can only be deciphered by a civilization's intellectuals who have acquired scientific knowledge equivalent to ours (we have only known about pulsars since 1967). But our remote friends will also need the blessing of an acute perspicacity: when Sagan displayed his graphic message to a number of astronomers, very few of them were able to make it out. But, according to von Hoerner's probabilistic equations, it is more than likely that any outer spacemen who behold Mrs. Sagan's unique exhibit will be far ahead of us, hence better equipped to construe its content than our earthling astronomers.

This Sagan-Drake embassy, however, has been harshly

criticized by a number of unappreciative colleagues. Despite its elaborations, some denigrators find it infantile. Others feel that the atomic symbols we use could well be indecipherable to other races, who could have an altogether different concept of the atom, with models more subtle than the electron, the proton and the theory of the quanta. After all, a proton and an electron are representations of our level of knowledge. All the same, on some planet similar to the Earth, specialists in elementary physics must have had recourse to such projections. Therefore, even if they have moved far beyond these concepts, they should be able to recognize them, just as astronomers today can recognize Ptolemy's chart of the sky.

Another point of dispute, somewhat carping, is the use of an arrow at the base of the design to indicate the engine's trajectory. The arrow, say the objectors, implies a civilization of hunters. Will outer spacemen interpret its meaning correctly? In any case, this is a matter of secondary importance.

One might well wonder whether a thorough examination of the craft would not yield more information to those humanoids who come upon it perhaps millions of years hence than the message does. The real substance of the Sagan-Drake communication, in some opinions, is the engine itself. From the structure of Pioneer X, extraterrestrial archaeologists would be able to cull a good deal of information on the civilization, by then surely extinct, that built it eons before, on its knowledge and the degree of technology it has acquired.

What are the chances that this exceptional example of Mrs. Sagan's art will ever reach the public for which it was intended? By all estimations, infinitesimal.

The first eventuality we must consider is whether Pioneer X will survive its long career in space without crashing into some astral body. The astronomers calculate this possibility on the same basis as that of a collision between two stars. Since astral bodies are so sparsely distributed in space, such an encounter is most improbable, something like one billionth in the arc of one billion years; and since our Galaxy is approximately 14 billions years old, the likelihood of a single collision would therefore hardly exceed one hundredth of a million. In ten million years, Pioneer X will hardly have one chance out of one hundred billion to penetrate another planetary system. If we figure one million advanced societies strewn throughout the Galaxy, as Sagan optimistically theorizes, then Pioneer X has less than one chance out of ten million billions of landing on some planet occupied by an intelligent race.

Despite this discouraging prospect, the vehicle could still fall into the hands of extra-terrestrials. If we exclude a landing on some inhabited globe, then the possibility remains that it could be intercepted in space by exponents of a supercivilization who can survey vast galactic regions, navigate through interstellar space and commandeer vagrant spatial matter of modest proportions.

Our present terrestrial techniques are far from permitting us to achieve such remarkable performances. We have not yet succeeded in exploring our own solar system down to the last detail, so that our knowledge of it is still incomplete. We do not know all about the bodies that gravitate on its periphery, or even whether there are definitely no more than nine planets in orbit around the Sun. We have counted only a fraction of the comets that seem to take form from the fringe of our original solar cloud; every year astronomers discover a dozen of them, and only when their capricious courses bring them close to Earth. Yet, we can foresee techniques that will enable us to survey a huge sector of relatively close galactic space, combining astronautics with a perfected radar system.

The authors of the message carried by Pioneer X are counting not so much on a delivery by means of a neat landing on some inhabited planet as on the chance that it will be retrieved somewhere in space.

Destination: The Stars

Astronautic technology has now developed to such a degree that man can almost certainly build aeroships capable of transporting passengers to Mars; but expeditions into realms beyond our solar system still remain a challenge we are unprepared to meet. At the maximum speed our spacecraft can travel, a manned vehicle directed to Proxima Centauri, our nearest neighbor star, would require 100,000 years to reach its destination. Powered by ionic or nuclear fuels, it would move much faster, but far from fast enough for any practical purposes.

On a commission from NASA several years ago, the Jet Propulsion Laboratory in Pasadena set about exploring the whole question of interstellar traffic, and eventually arrived at discouraging, if not wholly negative, conclusions. The reason for its glimmer of optimism lies with the discovery that the stars provide resources of energy which can be harnessed to propel a spacecraft through the interstellar void. This principle has already been demonstrated in practice: during its Jupiter flyby in December, 1973, Pioneer X drew energy from that planet's gravitation to step up its speed, and take on a supplementary velocity that will impel it into the Galaxy beyond our solar system.

The physicist Freeman Dyson has recommended exploiting the same gravitational effect of a double-star system, to be selected from among the densest of them—white dwarfs or neutron stars where matter has contracted. He estimated that, spurred on by this propulsion, an interstellar vessel could assume an added thrust to attain a speed of 1,200 miles per second, equivalent to 1/150th the speed of light.

Furthermore, he believes that at the present stage of our technological development, it should be possible to build vehicles that could skim through space at 6,000 miles per second after the first ten days aloft at the average acceleration of one G[1]. He proposed utilizing the shock reaction set off by nuclear explosions, which could be obtained by releasing H-bombs timed to explode half a mile behind the craft. Such a system could propel an engine at a speed one thousand times faster than any can now sustain. Even at such a rate, a spaceship racing for the nearest inhabited globe, thought to be some 18 light years away, would complete a voyage of one thousand years before it could return to Earth, with thirty generations of space crews at the controls. Interstellar navigation, therefore, is already possible in theory, but only at such restricted speeds that manned flights must be ruled out.

According to the theory of relativity, it is not unrealistic to envision astronauts traveling enormous distances even hundreds of thousands of light years, without aging very much—but only on condition that their vehicles move at a velocity close enough to the speed of light.

Let us imagine the case of a manned astroship powered by some technique yet to be discovered. Directed toward a given star, it travels at a constant acceleration of one G and once it has reached the halfway mark it slows down in the same manner. Calculations show that, following such a flight schedule, it will reach the nearest stars within a few years by the cosmonauts' watches, and the center of the Galaxy in 21 years. If it were to leave our Galaxy, it could penetrate the Andromeda Nebula in 28 years. But before the astronauts aboard set foot again on home soil, they would find the world totally unrecognizable; actually over three million years would have elapsed on Earth. This is because time aboard an astroship would flow at a different rhythm from time on Earth.

But a flight approaching the speed of light, so dear to the hearts of science fiction fantasists, brings up a number of fundamental difficulties. For one, even if some brilliant

[1]G: an acceleration equal to the acceleration of gravity, about 32.2 feet per second at sea level. It is used as a stress measurement for bodies undergoing acceleration.

physicist should hit on a way to transform the entire mass of motor fuel into energy, as Einstein's equation enjoins, an astroship bound for Andromeda would be obliged to carry aboard a load of combustibles 100,000 times heavier than its own weight. Moreover, total conversion of the fuel into energy is conceivable only if half of it were anti-matter. But no one has yet conjured up a magnetic container that could imprison this fugitive commodity.

There is still another barrier, apparently insurmountable. At the approximate speed of light, a collision with interstellar dust, or with the simple atoms dispersed in space, would cause an astroship to disintegrate. One possible preventive measure against such a mishap might be to equip the craft in front with a sort of twin-system fender, like a spatial snowplow, designed first to ionize oncoming particles, then to shunt them off the vehicle's path by means of a protective magnetic shield.

Some scientists predict that new theoretical breakthroughs in physics will point the way to the discovery of still unknown sources of energy. They theorize that future engines will transport men by means of new routes which will not traverse our space.

In some future era, science fiction will no longer claim a monopoly on space travel. Confirmed physics assures us that long-distance—indeed, *very* long-distance—intersidereal navigation is within the bounds of the possible. Significantly, American scientists appear to be more concerned with the prohibitive costs involved than they are with the technical difficulties blocking the way. As with the Apollo program, once the financial hurdles are cleared, these problems will somehow—and inevitably—find their own solution.

Under the auspices of the British Interplanetary Society, a team of technicians specialized in propulsion is studying the feasibility of building a space station that could fly to Barnard's Star, six light years away, to carry out a series of soundings. They are trying to determine whether, by using technologies now conceivable, it is possible to construct an "intelligent" automatic vessel that could control its own systems, navigate automatically on a long course, set its own objectives and adapt itself to an unknown environment. We know nothing about the planetary system escorting Barnard's Star other than the fact that the perturbations in the star's movements indicate two big planets—or possibly three—each with a mass comparable to that of Jupiter.

The Society's Project Dedalus is the brain child of a former Rolls Royce engineer, Alan Bond, propulsion specialist, and Anthony Martin, of the University of London. For eighteen months they inducted the team in the properties of an engine designed on the basis of forecastable techniques—possibly the intersidereal vessel of the XXIst century if, in the meantime, no unforeseen breakthroughs in propulsion developments bring to light physical resources as yet undiscovered.

Among the propulsion techniques suggested for interstellar voyages, the British technicians studied *Ramjet*, a method by which the motor would acquire energy from the hydrogen it collects on its travels; this would partially resolve the problem of the fuel's surcharge. But this combustible is sparse in the regions separating nearby stars.

Another propulsion possibility is photonic navigation, by which light is utilized to power a spacecraft from behind over long distances. Again, a sufficient pressure of radiations would be required. The laser, which makes it possible to avoid the dispersion of a luminous beam, offers such prospects. A beam of aligned light emitted from the terrestrial platform could accelerate a space vessel, even if its pressure is meager, because of the cumulative effect of a continuing action. This type of propulsion has been given serious attention, although powerful lasers which can function by continuous action are still pure science fiction.

The British technicians, however, believe that the most accessible and highest yielding propulsion system would work on the principle suggested by Dyson: an explosion motor impelled by small thermonuclear charges. The microbombs of deuterium and helium 3 would be projected at a ratio of 250 per second into a cusp-shaped magnetic field where they would be hit simultaneously by four high-powered electron beams and the plasma finally ejected at high velocity. The main engine would be employed until the vehicle's mass fell to a few thousand tons after which it would be jettisoned and replaced by a smaller reserve engine.

In space, any propulsion system dictates the mass of the vehicle. For a payload of 500 tons, Dedalus would require an engine hardware of 400-500 tons, plus the nuclear fuel—the microbombs enclosed at an extremely low temperature in a series of spheric reservoirs—which add up to some 150,000 tons. On its launching, then, the craft itself, its cargo and its nuclear fuel would weigh five hundred times as much as von Braun's mighty Saturn V and its convoy of sondes, which weighed 3,000 tons, including 2,600 tons of combustibles.

But the two systems are not comparable. Dedalus would not take off from the ground, but rather from a space station; and the pollution from its propulsion would be intolerable.

The first mission of the Dedalus propulsor would be to guide the vessel into a speed which would liberate it from terrestrial limitations, then free it from solar limitations to attain a cosmic velocity of forty-two kilometers per second. To arrive at Barnard's Star, Dedalus would have to maintain its propulsion regime for five years. At the end of this continuing acceleration, it must achieve a speed of 51,000 kilometers per second, or seventeen percent of the speed of light. The missile could then cruise for thirty-five years on its hyperbolic trajectory; there is no question that it could move in a straight line. Its navigational instruments must be geared to cope with any unexpected situation that might arise; its built-in logic would provide for its maintenance and take care of any defects that would crop up, making its own repairs. Such techniques are now being rapidly developed on today's spacecraft, but the length of Dedalus' voyage and its complex mission would require it to be a wholly independent unit, endowed with a lofty electronic intelligence quotient and broad faculties of adaptation during its peregrinations.

Once in the proximity of Barnard's Star, Dedalus would fly over that body without braking, much as Mariners have done in closer quarters. As it approaches, its automatic infrared telescopes should be able to single out the huge planets in orbit around the star. Its sondes would then take off, each in a different direction, to scour the system in search of other planets of the telluric type and their moons. These exploratory soundings would transmit their information to the main station, which would then relay it back to Earth. At the distance of six light years, these transmissions could not be intercepted by the radio antenna network NASA is now using to communicate with its interplanetary engines; its parabolic antennae measuring some 250 feet in diameter are utilizable only within the solar system. As we have said, they will lose all contact with Pioneer X once it has left our solar system behind. To maintain contact with Dedalus would necessitate a giant network of interconnected antennae 325 feet in diameter, such as the antenna now being studied for Project Cyclops by a team of NASA and Stanford University experts.

Dedalus, of course, will not be built, at least in the foreseeable future; too many problems remain to be cleared up first. But it is perhaps the prototype of exploration spacecraft of the coming century, when astronautics will be ready for it. To those who shrug it off as a futile dream, let us point out that the same British Interplanetary Society conceived a model for a Moon-landing apparatus thirty years before Armstrong and Aldrin were the first Earthlings to set foot on lunar soil.

For their part, the Soviet specialists have calculated the practicable flight trajectories of sondes launched in the direction of various stars among those within a range of twelve light years. The first of them would be equipped with ionic motors, in which the propelling gas would be created from ions accelerated by an electrostatic field. Of all the electric motors studied for spatial flights, this is undoubtedly the type most favored for extremely long voyages. Its propulsion would give it a weak but continuing, permanent thrust, and the gases ejected could enable the sonde to achieve a speed of tens, even hundreds of miles per second, by contrast with the 2.7 miles per second allowed by hydrogen, the best of all chemical carburants.

En route to Barnard's Star, in its first lap of 5.5 million miles, corresponding to the zone in which the terrestrial gravitational field is sensitive, the Soviet sonde would increase its acceleration to attain a velocity of 5,700 miles per second. This lap would require sixty years, after which the craft would cruise with its motor inactive for eighty percent of the voyage, and decelerate in the final phase. Because of the ionic motor's weak thrust, the braking would take thirty years. The total flight time from the Earth to Barnard's Star would last between 190 and 290 years.

Prehistoric Flight Bulletin

While the prospect of manned interstellar travel lies in abeyance for some time to come, the reverse side of the coin is the subject of intriguing speculation:

Have *they* visited *us*?

A widely diffused literature maintains that strange astroships from other worlds have indeed landed extraterrestrials on our soil. Such affirmations, however, are based not on demonstrative analysis but rather on fragmentary information intended to arouse public curiosity—and sometimes succeeding—but often at the cost of exaggerations, omissions that leave only half-truths and deliberate manipulations of documents quoted.

Turning back to early history, we find numerous legends, often popularized by religious leaders, that tell of heralds sent from heaven, of giants and semigods who trod the Earth before the dawn of history. It is tempting to attribute these tales to surviving figments of distorted memory, which dimly recall living creatures from some other world or worlds, who once settled on our planet. These visitors from the firmament presumably taught their backward hosts the rudiments of highly complicated techniques, such as primitive man had never dreamed of before, and showed them how to construct great monuments, which archaeologists today are hard put to ascribe to the ignorant, prehistoric ancestors of the peoples who have inherited them. Surviving relics in two geographical sectors are especially vulnerable to similar exegeses: the Mexican-Andean cultures in the New World and the Near East Sumer civilization in the lower Euphrates Valley and the great cultures which succeeded it, including Judeo-Christianity. Another example is Easter Island, with its gigantic stone figures of an origin that defies all efforts to clarify the mystery. But such phenomena exist in many places throughout the world, and some individuals versed in this lore perceive tangible traces of extra-terrestrial supermen in the mythologies of the Eskimos, the Saharan tribes, the East Indians, the Chinese and other ancient peoples.

Today, these explications are no longer acceptable, and all such legends are subject to altogether different interpretations. In every part of the world, the historians of religion find parallel traditions dating back to remote ages when superior beings—gods, heralds of heaven and light—often waged war against the evil powers of shadowy figures who dwelt in the darkness, in the abysses and the entrails of the globe. Everywhere they detect vestiges of this primitive will to identify gods with the forces of life, to behold them as messengers of light. But until concrete evidence yields incontrovertible proof that supermen did indeed alight on this planet from their spacecraft, we must assume that the great archaic myths were invented by authoritarian leaders to play upon the superstitions of their gullible folk by summoning up cosmic powers as an instrument easily wielded either to terrorize or reassure them.

Furthermore, the very multitude of great monuments existing all over the globe makes it hardly credible that they were left behind by super humanoids who came from the blue. An invasion on such a massive scale could not have occurred without our visitors leaving some tangible traces of their sojourn on Earth. If archaeologists have dug up skeletons of giant prehistoric races, there is nothing to indicate that they were migrants from other worlds. No remaining graphic document or monument coming down to us from cultures that flourished in the dim recesses of history hints at contemporary cosmonauts. To the ethnologist, the Mexican slab at Palenque (site of a once great Mayan metropolis), reputed to portray a Mayan astronaut on his spaceship, is no more than a metaphysical allegory; there is nothing mechanical about it. Likewise, experts see the head pieces of pre-Colombian and Saharan diving suits as masks, ritual ornaments or metaphysical symbols rather than the trappings of supermen from the sky. If we were to accept these fantasies as authentic, then we should also have to name Rome, with its hundreds of churches, the world's capital of extra-terrestrial astronauts, and accept its large population of haloed saints as cosmonauts who flew to Earth from heaven-knows-where.

Flying is an age-old, worldwide dream, which the vision of birds on the wing has never ceased to stir in the imagination of men everywhere. The Greek myth of Icarus, whose waxen wings melted in the heat of the Sun, is akin to the air-borne carpets of Persia, the flying chariots in the sacred books of ancient India, and the subjects of innumerable ancient Chinese drawings. Here again, extra-terrestrial astronauts are wholly extraneous.

The first scientific enquiry into this area began in 1959, when Professor Agrest, a Soviet physicist, conducted a research project to weigh the pragmatic values of myths as history. He did not exclude the possibility that some of them, particularly the very early poems composed by Middle Eastern bards some 4,000 years ago, could have incorporated mnemonic remnants of a visit to Earth by outer spacemen.

A popular Sumerian legend told of superior specimens who rose up out of the Persian Gulf waters, mingled with the local inhabitants and instructed them in the elements of technics and the sciences. Sagan, who extended his investigations into the realm of myth as fact, noted the abrupt metamorphosis, some five thousand years ago, of the Sumerian people from a very rustic culture into a society suddenly adept in sophisticated technics and sciences, particularly mathematics and astronomy.

The Biblical legends, with their Babylonian roots, readily lend themselves to interpretations admitting of spacemen. J. Robert Oppenheimer, who directed the Manhattan research project which led to the atomic bomb, once observed that the fall of Sodom and Gomorrah seemed very like the consequences of a nuclear explosion; and the various accounts of supranatural kidnappings, such as the sequestration of Enoch and Ezekiel as well as the innumerable episodes in which winged angels took part, could be attributed to astronauts from outer space.

Sagan carried out a small statistical study on the probable chances that the Earth might have been visited by extra-terrestrials. His extrapolation, of course, was certainly flimsy, and his points of departure highly arbitrary, yet the method he used was interesting. He began by estimating the probable frequency of direct contacts among hypothetical advanced societies in the Galaxy according to von Hoerner's system, then attempted to deduce, from a theoretical calendar of exploration, the frequency of the visits extra-terrestrials presumably paid to the Earth, or rather to our solar system.

The problem was not unlike the dilemma of a man with a mathematical turn of mind who, shipwrecked on a desert island, tries to calculate his chances of sighting a rescue ship. He takes note of the volume of traffic in the surrounding sea, figures how far away the islands nearest to his might be, and lists the reasons why anyone would want to embark on a voyage to his speck of land. Ultimately this solitary Robinson Crusoe reckons that he may have to wait as long as ten thousand years before the next visitors land on his shores. This is the same mathematical conclusion Sagan reached.

Pointing out the insubstantial nature of Sagan's numerical bases, Shklovsky pronounced it a "probability difference of zero" that outer spacemen ever landed on Earth in prehistoric times.

A CHALLENGE TO SCIENCE

The UFO Conundrum

Thus, Sagan and Shklovsky posed the question of visitors from outer space in past ages; but does it apply to the present as well? Is it within reason to suppose that vehicles from other planets are coming to call—or spy—on us even now, in the twentieth century? What are the true facts that lie behind the mass of data compiled on flying saucers? For the present, the scientists are trapped in a dilemma of embarrassment.

Their quandary arises not so much from the nature of the problem as from the particularly awkward circumstances that impede any rational, systematic effort to study it. Unfortunately, flying saucers—and, more generally, unidentified flying objects—are exasperatingly elusive, random apparitions scientists must study at second hand from testimonies which are more often than not sketchy or of dubious validity.

In point of fact, flying saucers, or the alleged phenomenon as such, represent only a small fraction of the unaccountable flying objects we know about. These objects, after attentive examination, can prove to be airplanes, helicopters, test balloons or space vehicles returning to the Earth's atmosphere. They can be meteorites or atmospheric conditions, such as tornadoes, inversions of temperature, lightning bolts, formations of ice crystals, decoys sent aloft by defense services, metallic anti-radar ribbons or various other matter. Beacon lights playing on low-lying clouds and the radiations of exceptionally brilliant celestial bodies, like the planet Venus, can create optical illusions. Any air-borne matter in motion, or seemingly in motion, to which a definite label cannot be assigned, is automatically classified as an unidentified flying object.

Competent enquiry will eventually attach a label of one sort or another to most of these illusions; but what about those other observations that defy all efforts on the part of the specialists to determine exactly what they are? Either the specialists must assume that such sightings will eventually yield to some natural explanation not yet known to science, or else concede that they have something to do with artificial flying constructions piloted by living creatures. There are no other alternatives.

When UFOs appear to be devices fabricated by intelligence not of this world, the validity of the testimonies given by the observers will be put to a merciless third degree. Most of the probers will insist on the likelihood of fraud, collective hallucination, schizophrenia or downright simple-mindedness. The witnesses, therefore, are put under a rigid examination, not only by experts in the physical sciences, but also by psychologists.

All the same, numerous UFOs descried by human eyes have also been recorded simultaneously by photography, radar or some other means of detection. And many of those still classified as unknown objects have been reported not by publicity seekers or feeble-minded souls but actually by airplane pilots, scientists, especially astronomers, and radar technicians.

A thorough, impartial analysis of their observations has been hampered by the stubborn reserve on the part of many scientists who have adopted an attitude of extreme skepticism for fear of finding themselves the target of public ridicule. Their reticence has certainly been stiffened by the rash pronouncements coming from persons who champion the flying saucer theory without possessing the qualifications to do so, by an upsurge of phony flying saucer literature and an outbreak, particularly in the 1950s, of innumerable parascientific flying saucer societies aroused to action by that literature, but lacking the competence to shed any real light on the mystery.

Outside the United States, no country has even sponsored a public UFO investigation, although everywhere private organizations have taken steps to affiliate into federations and pool their findings; but again, unfortunately, they lack the elaborate, sophisticated equipment they would need to achieve significant results on their own. Nevertheless, they occasionally compile useful information, helpful to a small number of scientists—astronomers and specialists in the physics of the atmosphere—who are privately giving serious attention to the problem.

Certainly the majority of these men share the belief that extra-terrestrials are holding the Earth under surveillance, and agree with the late American scientist James E. McDonald, who adjudged the possibility that some UFOs are manned spaceships as "the least mediocre hypothesis."

Ronald Bracewell, an Australian astronomer, considered it likely that outer space scientists could have combined automatic aeronautics effectively with hertzian contacts in the reconnaissance relay sondes they have most probably launched into space. Galactic civilizations, even those in an advanced age of communications, would have little chance, he thought, to establish long-distance communications which are possible on the interstellar scale only by means of equipment so gigantic that more than likely they do not even exist. Therefore, the theory of relay sondes would seem to be the most logical.

Such a perspective, however uncybernetic it may be, profoundly alters the whole face of galactic communications. In proposing automatic relay communications directed to home bases from the Earth's atmosphere, Bracewell, in a way, endorsed the concept of flying saucers. But if there is any chance that his supposition might be correct, then one might well ask a leading question: why have the world's detection instruments, highly developed as they are, never tracked the movements of relay sondes in the environs of the Earth? Bracewell, a specialist in such techniques, has a ready reply. He reminds us that extremely powerful radio emissions coming to us from within our own solar system, such as Jupiter's radio-electric radiations, went totally ignored for a very long time.

The U. S. Air Force Investigates

Unidentified flying objects became a source of worldwide fascination and alarm only at the end of World War II; yet, leaving aside the prehistoric and Biblical legends of gods, demigods and supermen landing on Earth from divine regions in outer space, the number of UFO sightings recorded all through history since the most ancient times will overwhelm anyone who can take the time and trouble to gather the formidable mass of pertinent data to be found in every quarter of the globe. Why, then, has this phenomenon come so forcefully to the world's attention only in the past few decades?

For one reason, Earth dwellers have never before been mentally conditioned to accept the notion of living beings on other planets. Ever since the second century A. D., when the astronomer Ptolemy proclaimed his Ptolemaic system, which presumed that the Earth was the center of the Universe and the only inhabited globe, all mankind subscribed to it. The Church upheld his theory, and Church dogma was not to be questioned. Actually, in the past millennia, many UFO witnesses, bound by religious precepts, were unwilling to believe what they saw. With Kepler's development of the Copernican system in the seventeenth century, the Ptolemaic theory was demolished; but it was not until the dawn of the space age, with the amazing revelations of advancing astronomy and the discoveries of astronautics, that earthlings began to wonder whether our planet was indeed the only one in the cosmos to sponsor intelligent life.

Another reason was, of course, the lack of mass media to flash news of UFO observations around the world. Communities were isolated, and communications poor. Word of sightings seldom traveled very far beyond the confines of the areas in which they occurred.

Nevertheless, a study of UFO incidents to be dug up in local registers will show that an untold number of them have taken place all through recorded history, many similar to those that have made headlines in the recent past. Often in bewildered disbelief, individuals at all points of the compass told of seeing strange objects sailing through the sky. Their shapes were variously described as discs, spheres, "cloudships," cigars, wheels, saucers, eggs, submarines, even chimneys and fountain pens. Some were said to move silently while others made roaring noises that blasted the eardrums. Not a few of them stopped and hung motionless in mid-air for varying lengths of time, or landed on the ground. Sometimes the occupants were seen to emerge from their craft. At night these flying constructions emitted soft or blinding lights—white, red, orange, green, blue—long before electricity was ever dreamed of; unable to define such unwonted illuminations in more accurate terms, witnesses spoke of them as "fire."

Ancient Egyptians recorded sightings in hieroglyphics on papyrus. Citizens of the far-flung Roman Empire clutched at their togas at the sight of air-borne vehicles unlike any chariot that raced in the Circus Maximus, and far swifter. Pliny wrote about "shields of fire" (*clipei ardentes*), and Titus Livy noted that "phantom vessels appeared in the sky" (HISTORY OF ROME, XXI and XXII). During the Middle Ages, such apparitions were commonly thought to be symbolic religious manifestations sent from heaven. In India, a Sanskrit text spoke of "discs that roared in the sky." After the great age of explorations and the discovery of the Americas, sightings were recorded all over the expanding known world, some by astronomers and observatories, where the reports are still preserved, available to anyone who might want to see them. In 1878, a Texas farmer saw a disc high above traveling at an incredible speed; he likened it to a saucer, a metaphor that would become a household term over half a century later. Similar observations continued through the present century, notably at Fatima in Portugal where, on 22 January 1922, some seventy thousand persons gazed as one at a vision overhead; the great majority believed it to be an image of the Holy Virgin, but a few thought otherwise. During World War II, warplane pilots saw mysterious apparitions during their flights which they could not identify. These may or may not have been illusions projected by the new combat technologies to which they were unaccustomed.

Beginning in 1946, with the emergence of a new space consciousness no longer limited to a few scientists but now taking root in the public mind, the UFO phenomenon began to take on a new dimension. In that year, a wave of UFOs overflew Sweden; subsequently, similar waves were reported elsewhere over Europe, particularly France. In the United States, the first sensational postwar alert occurred in June 1947, when Kenneth Arnold, an American businessman piloting his private plane over Mount Rainier in the State of Washington, radioed to ground that he had unmistakably seen a number of flat, round, silver discs cruising the sky in formation. The news of his momentous experience shook the country into a state of collective awareness that something strange and ominous was going on, a possible threat that demanded urgent investigation. The incident became a classic in flying saucer annals.

Other similar reports began to emanate from all parts of the country. The United States Air Force, entrusted with the solemn mission of keeping the nation's air space under strict surveillance to protect national security, hastened to initiate an enquiry into these curious apparitions. Its main preoccupation was that these potentially dangerous flying objects might be hostile espionage devices of Soviet origin.

The cold war, the development of international strategic missiles and the new science of astronautics, with its fearsome implications as well as its blessings, prompted the Air Force to blanket the country with an extensive network of space surveillance. But the Early Warning System, on the lookout for unfriendly intercontinental rockets, failed to detect any UFOs, which never seemed to cross the nation's borders. They were invariably local affairs, fugitive spectres that vanished like soap bubbles after a brief appearance over one area or another. The scientists assigned to investigate and technicians in the informational services were dubious.

Beginning in the early 1950s, however, flying saucers provided a thrilling, unique literary subject outrageously exploited in potboilers that met with an astounding public success, thus swelling many a happy author's bank account to unprecedented figures. A book entitled *Flying Saucers, Serious Business* rang up sales of some 1,300,000 copies. The text, by radio reporter Frank Edwards, simply anthologized a number of press clippings and second- or third-hand reports by witnesses, overwhelming the reader by the sheer weight of these testimonies. The author assumed the heroic guise of a St. George slaying the Air Force dragon and the secret security agencies, which he accused of attempting to conceal the true nature of the flying objects.

That this genre of literature triggered a nation-wide obsession with UFOs was illustrated by Dr. Edward Condon, who presided over the first civil commission of enquiry into these phenomena from 1966 to 1968, when he reported that within those two years, some forty thousand American school children wrote in to the Air Force to ask for more detailed information on them.

At the end of 1947, Secretary of Defense James Forrestal set up the first special study commission called Project Sign, or more familiarly, Project Saucer. One week after its inception, the members were confronted with an incident that shocked the nation. On January 7, 1948, a flying object was signalled over Madisonville, Kentucky, ostensibly traveling toward Fort Knox. Officers on the ground could see it clearly with the naked eye; the observatory at Goodman Base tracked its movements. A patrol of three interceptor planes took off to chase it. An hour after the alert, one of the pilots, Captain Thomas Mantell, radioed back that the ship was hovering directly over him. "The thing is enormous!" he gasped. "I'm going closer to see it better." But then the spaceship swooped sharply upward. The other two pilots lost sight of Capt. Mantell when his plane disappeared into a fog bank, gave up and returned home. Half an hour later, Mantell notified his base that he was still climbing but that he would have to abandon the chase at twenty thousand feet because he had no oxygen tank aboard. Forty-five minutes later, fragments of his wrecked plane scattered over an area of several square miles. His body was never found.

Among other incidents involving professional aviators early in the Project Sign enquiry was one reported by the pilot of an Eastern Air Lines DC-3 in flight not far from Montgomery, Alabama, when a cigar-shaped UFO passed extremely close to his craft "with its porthole lights blazing," and another on October 1, 1948, when a military pilot, Lieutenant George F. Gorman, of the North Dakota National Air Guard, found himself pitted in an "aerial duel" with a UFO over North Dakota.

Basing its conclusions on these and innumerable other sightings, toward the end of 1948 the Project Sign commission reported its findings to Air Force headquarters: UFOs were not of Soviet but rather extra-terrestrial origin. The Pentagon demurred and demanded further proof of this outlandish allegation but, dissatisfied with the commission's final report, the Department of Defense rejected it and called off the enquiry.

The following year, the Air Force set up a second commission, Project Grudge, which confined its duties to little more than receiving and classifying sighting reports. Now the UFO enquiry was reduced to the task of assessing the credibility of witnesses subjected to rigid examination by the unit's psychological section, while the number of sightings which could not be explained were arbitrarily relegated to a negligible margin. Clearly, Project Grudge aimed to avert the risk of a national panic sparked by a widespread flying saucer mania. But in its effort to stave off a wave of public hysteria, the commission left no room for an impartial scrutiny of the situation.

With sighting reports still coming in with unabated persistence, in September 1951, the Air Force information service once again set about to analyze them. Appointed to head the revived program, Captain Edward Ruppelt hired technicians and scientists and commandeered a computer, an electronic card-index system and other pertinent equipment; but his lesser rank indicated that the enquiry was a matter of no serious consequence to the Air Force hierarchy.

The following January, a number of former Air Force men who refused to accept the air arm's official attitude on the issue and questioned its methods of procedure, banded together to found a private investigatory body, the Aerial Phenomena Research Organization, for the purpose of putting incoming reports on both explicable and inexplicable cases to more thorough analysis and releasing their findings publicly in a bulletin.

Later in the same year, the Air Force had another change of heart and replaced Project Grudge with the more serious Project Blue Book, retaining Captain Ruppelt at the helm. "The people at the Air Technical Intelligence Center," he would later write, "decided that the time had come to make an estimate of the situation. The situation was that of the UFOs; the estimate was that they were interplanetary objects . . . The dossier was a thick one, containing the Air Force's analysis of many . . . incidents. All of them had been reported by scientists, pilots and other equally credible observers, and each case was an unknown quantity."

The new unit was authorized to function more freely and on a broader scale, maintaining direct contacts with all air bases, astronomy observatories, meteorological stations and other installations. Air Force flying crews and scientific observers were alerted and asked to relay all sightings of a suspicious nature.

On 26 July, two radar stations at Washington's National Airport as well as its control tower spotted eight UFOs over the area. The sighting was confirmed by a third radar station at the nearby Andrews military base, which reported that one of the flying objects abruptly broke ranks and flew off at a speed of some six thousand miles an hour. At dawn, some observers clearly saw a huge luminous orange sphere overhead. Intercept planes took off to pursue it, but the mysterious object promptly vanished. When it suddenly reappeared, another chase proved to be no less futile. Meanwhile, passenger aircraft pilots signalled that they, too, had seen strange lights in the very areas indicated by the observers, but could not identify them.

Oddly enough, other observatories and lookout posts in Washington recorded nothing of these incidents. But immediately afterward, another alert was sounded when similar apparitions were spotted over Langley Field in Virginia, over California and Michigan at approximately the same time. Was there any connection between the two events?

Alarmed, General John Samford, chief of the Air Force's information service, hastily summoned a meeting of his aides, during which the conferees agreed to set up still another enquiry commission of experts. Yet, masking his feelings behind a cool front, the General calmly notified the press conference that "UFOs are nothing more than weather phenomena;" and an Air Force communiqué announced that neither the Army, the Navy nor the Atomic Energy Commission were carrying out experiments that had anything whatever to do with these flying objects, emphasizing the fact that UFOs were not phenomena or any kind of apparatus produced by the Army or the Atomic Energy Commission. "We do not know the causes of the phenomena signalled," the communiqué read, "but we are convinced that they constitute no menace of any kind to our country. We cannot, of course, assert that they are not visitors from other planets or from some foreign country, but we are of the opinion that they are physical phenomena we are unable to explain at our present stage of scientific knowledge."

Thus, in its effort to forestall nationwide consternation, the Air Force, caught in a paradox, was obliged to confess that it was still in the dark in the matter of UFOs.

Dissension

The news of the sighting over Washington's airport aroused such a strong reaction throughout the country that the Air Force brass were obliged to counter it with some rational explanation of these mystifying visions that could travel at incredible speeds and suddenly vanish when any attempt was made to track them down.

Bowing to the pressure, the Air Force paid lip service to America's time-honored tradition of open public information by launching an investigation deliberately designed to mislead the public. But it was unprepared to cope with the mass of incoming reports on UFO cases that could not be dissembled to fit neatly into its scheme, and it was exasperated by flying saucer enthusiasts who exposed the Air Force spokesmen for faulty reasoning, sometimes catching them out *in delictu flagrante* when they perpetrated gross contradictions.

By the end of 1952, the matter became so pressing that the White House released a statement in defense of the Air Force, upholding its categorical denial that such aerial phenomena had anything to do with secret weapons, missiles or new types of warplanes under experimentation in the United States; and a year later, President Eisenhower, offering his paternal comfort to anxious voters, announced that "flying saucers exist only in the imagination of those who see them."

When Dwight D. Eisenhower assumed the Presidency in 1953, from the outset of his Republican regime, the authorities were clearly determined to gag those troublemakers who saw flying saucers as possibly something more than figments of overactive imaginations. In January of that year, the new commission of experts, set up after the Washington airport scare, met at the Pentagon to decide what course of action to follow. Chaired by H. P. Robertson, a physicist from the California Institute of Technology who often doubled as scientific consultant to the Department of Defense, the unit consisted of other prominent scientists, including Luis W. Alvarez, Nobel Prize winner in 1968 in physics, as well as the director of the Air Technical Intelligence Center and three representatives from the C.I.A. In its final statement, the commission reiterated the claim that UFOs were nothing more than natural atmospheric configurations. Some observers believe that strong pressure from the C.I.A. and the A.T.I.C. might have been a decisive factor in the commission's action.

Thus, Project Blue Book went into decline. Its chief, Captain Ruppelt, resigned his post, to be replaced by a G. I., a sharp downgrading that seemingly marked its imminent demise.

Forthright federal measures to discount the whole issue of flying saucers were not long in coming. In the summer of 1953, an order from the Undersecretary of Air stipulated that the percentage of inexplicable cases made known to the public must be pared to the minimum. Sighting reports would henceforth be turned over to the Air Defense Command, either because the aviation authorities wanted to convince the public by its own methods that all sightings were insignificant and harmless or hush up the whole issue completely.

The Eisenhower administration abetted them with strongarm actions. In the fall of 1953, two private flying saucer organizations, the International Flying Saucer Bureau and the Civilian Saucer Investigation, were summarily suppressed on the strength of a 1952 regulation which reserved the gathering and dissemination of UFO information as the exclusive province of the military. Thus ended, presumably, all UFO speculations that failed to hew to official interpretation. Henceforth, flying saucers, like good Republicans, must respect the accepted laws of physics. Those that resisted would end up in the forgotten files of the Air Force's intelligence service. Nevertheless, the Air Force was not altogether indifferent to the problems or wholly remiss. Taking due precautions, it covertly equipped its main bases on both domestic and foreign territories with spectroscopic cameras, so that its technicians could analyze the luminous spectrum of any eventual, unwelcome visitors from outer space.

These stringent measures against nonmilitary dissemination of UFO data, which openly defied the American constitutional right of free speech and press, were adopted in part to minimize the effects of the fantasies allegedly experienced by another writer, one George Adamski, whose books on UFOs were piling up huge sales in the United States. Adamsky claimed that he had actually seen flying saucers land on Earth and had chatted amiably with their crews.

In 1955, the Air Force published a decree regulating its new methods of investigation. Entitled *Report 4* it required all sightings classified as inexplicable to be turned over to its Psychological Section, confirming the air arm's intention of reducing the number of controversial cases to the very minimum.

Meanwhile, some of the technicians in the information service and other members of the Air Force fired off a new protest against these cover-up tactics. In 1956, this disaffection with official policy led to the foundation of still another private investigating group, the National Investigating Committee of Aerial Phenomena, with a membership of several thousand persons, mainly Air Force officers, also including a retired admiral and a former C.I.A. division chief, in open defiance of the 1952 regulation.

During the same year, Edward Ruppelt published a *Report on Unidentified Flying Objects*, in which he attacked the Air Force's UFO policy and procedures. In a counter-offensive, a spokesman for the Pentagon, Lieutenant-Colonel Lawrence J. Tacker, denied his accusations in a book entitled *Flying Saucers and the U. S. Air Force*, hotly defending the authorities' actions and denouncing their critics, who were agitating for a public exposé of the unvarnished truth.

In a manifestation of exasperation, the National Investigating Committee and other private groups called upon Congress to authorize a counter-investigation, but no response to their plea was forthcoming from Capitol Hill. The most insistent demand came from J. Allen Hynek, a scientific consultant to the Air Force, who had published an exhaustive review of the situation in the April 1963 issue of the Yale Scientific Review, dwelling on the shortcomings of the Air Force enquiry and appealing for a thorough exploration of the question by qualified, unbiased scientists.

Hynek's Dilemma

Director of an astronomical research institute in Illinois, J. Allen Hynek also acted as consultant to the U. S. Air Force in its UFO enquiry. In this latter capacity, he reviewed thousands of cases and investigated a good number of them himself. His instructions were to report only those cases that could be disposed of as common astronomical occurrences—meteorites, planets, stars in alignment, and the like.

For a while, Hynek dismissed flying saucers as a trivial

matter, hardly worth the trouble of taking any serious official action. But as he studied more and more cases, he gradually changed his mind. A considerable number of them, he saw, could find no solution within the limits of mankind's scientific knowledge.

In his *Yale Scientific Review* essay, Hynek dwelt on the negative effects brought about by the investigators' prompt willingness to doubt the word of the witnesses they examined. Contrary to widely held opinion, especially as commented in the press, most of the unresolved cases had been sighted not by persons wholly untutored in science or bereft of a respectable intelligence quotient; indeed, the average mentality of the witnesses interviewed, Hynek pointed out, was higher than average. Some of the most convincing among them were well grounded in science. Thus, he listed four reports from astronomers on duty at their observatories when the events they described took place, five from specialized technicians, and one from the vice-director of "one of our foremost national technical laboratories." All except three told of brilliantly illuminated spacecraft maneuvering in the air.

His remarks were fully corroborated by Captain Ruppelt who, during his office as director of the Air Force enquiry commission, had quietly undertaken a confidential poll of various American astronomers; of those queried, eleven percent admitted to having themselves seen UFOs they could not identify.

Despite his affiliation with the Air Force, Hynek held firmly to his unorthodox views. To be sure, he allowed the air arm due credit for the vast quantity of information it had so diligently gathered; nevertheless, in his opinion, "no really scientific examination of the UFO phenomenon has ever been undertaken, in spite of the enormous volume of data available."

Others agreed. In 1964, the National Investigating Committee of Aerial Phenomena published its *UFO Evidence*, a general report on what was being done—and not done—about UFOs in the United States. Flying saucer societies here and there throughout the country began to step up their activities as new observation reports poured in. The commercial air lines, preoccupied for the safety of their passengers, joined the movement when sixty-seven carriers lent their support to a new civilian Volunteer Flight Officer Network, with a membership of passenger aircraft pilots who agreed to report any strange flying objects they might encounter in the air to the Smithsonian Astrophysical Observatory in Washington, the International Unidentified Objects Committee, a private unit headed by Herman Oberth, a pioneer in astronautics or, later, to the Government-sponsored Condon Commission. The Smithsonian Observatory alone received several hundred reports.

Prodded by these developments, Congress began to stir. Speaker of the House John W. McCormack declared emphatically, "It is my belief that the Air Force is not telling all it knows about unidentified flying objects. We cannot disregard so many unassailable witness reports." Congressional leaders summoned experts for a consultation. Representing the Air Force, Major Hektor Quintanilla limited his comments to a flat statement that no such things as flying saucers existed. But Hynek, in his rebuttal, cited the pronouncements of serious-minded scientists, warned that an upsurge of UFO activities could well exacerbate a dangerous situation, and reiterated the arguments he had set forth in his *Yale Scientific Review* report for those solons who were unacquainted with it. In conclusion he demanded that "the Air Force must be relieved of this mission. The enquiry must be turned over to a civil commission composed exclusively of authentic scientists."

In the August 1966 issue of *Science* Magazine, Hynek published an appeal to the whole American scientific community to bestir itself, and reproached its members for ignoring the gravity of the UFO quandary, which posed a serious challenge to the entire profession. Reminding his readers that a good many sighting reports were altogether credible, and scoffing at the official dread of public mystification and hysteria which, he wrote, were much less in evidence than the Air Force assumed, he offered to put his experience and all his files at the disposal of those among his colleagues who were willing to study the problem with an open mind. His plea ended on a note of disillusionment: "I cannot dismiss the UFO phenomenon with a shrug of the shoulders. I have come to believe that there is a tendency in twentiety-century science to overlook the fact that there will be a science of the twenty-first century."

Late in August of 1966, a curious incident took place at a North Dakota Air Force base for intercontinental strategic missiles. In the underground radio room, the officer on duty saw that his set had suddenly gone dead, even though it was equipped with a multiple system which should have prevented any such breakdown. At the same moment, an officer on the ground spotted a UFO high overhead, aglow with a red light. At about the same time, the base's radar screen showed an object flying directly over the area at an altitude of 100,000 feet. The object descended to the ground, rose again and disappeared. The coincidence of the ruptured radio communications, visual observation and radar blips has never been explained.

Alarmed by this invasion of its highly sensitive air space, the Air Force took measures to reinforce its surveillance of the skies, and new regulations were put into effect during the month of September. The official brass even consented to open its UFO files to scientists, but this proved to be only a temporary change of heart; the files were subsequently withdrawn from outside inspection and once again shrouded in secrecy.

Lyndon B. Johnson, then President, demanded that "all light be shed" on the matter, and the enquiry commission Hynek had been agitating for was finally announced. The mission was now entrusted to a group of the University of Colorado in Boulder, under the guidance of Professor Edward U. Condon, a physicist. The university enjoyed the unique advantage of its location, not far from the

National Center of Atmospheric Research and the Environmental Research Institute.

Encouraged to carry out its probe in depth "without interference from the Air Force or any control on its part," the newly constituted Condon Commission was staffed by physicists, astrophysicists, among them Hynek, phychologists specialized in behaviorism, a meteorologist —all university professors—and two graduate students in psychology and another in literature. To their number were added some one hundred researchers.

Satisfied that his recommendations had been heeded, Hynek was soon to discover, however, that the work of the commission was taking a direction he had not expected. Indeed, the true nature of its intentions came to light when two researchers, one of them a full member of the group, were summarily dismissed for releasing a secret document written by Robert Low, coordinator of Project Colorado and Condon's foremost collaborator. Low's paper stated in no equivocal terms that the commission would liquidate the UFO issue once and for all by giving top priority to false sightings reported by persons of doubtful mental stability, and to other observations mistaken for unnatural phenomena. Thus, by pre-arrangement, the work of the group was not aimed at a bona fide enquiry but rather at an anti-flying saucer campaign, its effectiveness buttressed by the persuasive authority of the eminent men taking part—a red herring operation slated to follow in the footsteps of the Robertson enquiry, but on a more spectacular scale.

It is possible that Low, a member of the university's high command, had drawn up this document—in August 1966, before the commission was formed—to bolster Condon's position with the incumbent Air Force Secretary. The chase after federal contracts for research projects is a popular sport with the great American universities which pattern their aggressive tactics after the practices of big business, and resort to every means of wooing a potential client with glowing promises of satisfying his wishes in return for a lucrative compensation. In this case, aside from the prestige and nationwide publicity won by the fortunate university chosen to execute this much-coveted research mission, the Air Force contract offered the alluring prize of $200,000, later increased to $530,000.

The Condon Commission lasted until the summer of 1968, when it sent its final report—which Hynek refused to sign—to the National Academy of Sciences, which promptly approved it, with no questions asked. The wholly negative findings repeated the same old refrain: there were no such things as flying saucers, therefore no need to go to the trouble and expense of mobilizing the nation's scientific apparatus for a nonexistent problem. While certain aspects of the UFO phenomenon still remained to be clarified, the report admitted, proof that they were all of natural origin would be sure to come to light eventually. Writing in the *Bulletin of Atomic Scientists*, in December 1969, Condon made no reference to these obscure "aspects," but rather confined his comments to the frauds and errors of identification, which he alleged to be the inspiration of the "flying saucer cult"—an "astronomical imposture," he called it.

After conducting its UFO investigation program for twenty-two years, the U.S. Air Force struck it off the books at the end of 1969, in the process dispensing with Hynek's services. The probe, however, is not altogether defunct; sighting reports are now funnelled into the Foreign Technology Division, a military information agency specializing in the study of advanced technologies.

Phenomena That Defy Investigation

The Condon Commission's final verdict was by no means unanimously approved by all its members; there was also a minority report, which Condon managed successfully to keep under wraps. Meanwhile, by upholding the majority's conclusions, the National Academy of Sciences blocked any possible recourse to Congress, the only body empowered to authorize a new investigation. Thus, the only federal organizations left to study the UFO mystery were units of the secret services.

In the United States, the National Academy of Sciences is considered the nation's most trustworthy guarantee of scientific competence. This venerable, if fusty, institution is not a repository of ideas; instead, its function is mainly to define priorities for proposed research programs. In 1963, its Commission on Space Sciences advocated that NASA's explorations into outer space life should be ranked as a high priority scientific objective. This recommendation was a mark of the Academy's esteem for the remarkable results NASA had achieved in its interplanetary soundings at relatively moderate costs. To fulfill and carry forward its Academy-approved mission of specialized research into exobiology, NASA was able to maintain its center at Ames, California, at a high level of activity—and quality. Thus, the National Academy's towering authority lent considerable weight to the Condon report by virtue of its confirmation.

During the twenty-two years the Air Force pursued its semi-public enquiry into UFOs—from 1947 to 1969—its specialists examined some twelve thousand cases of sightings. Dr. Hynek, who, as adviser to the program, had had access to its files, avowed that over three thousand of them were totally inscrutable; they still remain unidentified flying objects today.

Of this number, the Condon report analyzed over ninety-five. As Hynek pointed out, every Commission member was delegated to study cases within his particular sector of the enquiry—including the scrutiny of photographs, radar signals and events that had left significant traces in their wake—which could not be discarded as false. Hynek charged Condon with conducting not a single investigation himself and dedicating most of his final report to the exposure of hoaxes and trivial errors of identification.

As its end result, the report virtually left the same residue of unaccountable sightings that had defied the Air Force investigators for over two decades. Indeed, its percentage of unresolved cases appears to be even higher than the corresponding percentage in the Air Force files.

One Commission specialist, Professor Hartmann, a University of Arizona astronomer, summed up his conclusions in these words: "The present data are compatible with but do not establish either the hypothesis that the entire UFO phenomenon is a product of mis-identification, poor reporting and fabrication, or that a very small part of the UFO phenomenon involves extraordinary events."

Other observations in the majority report contradict its overall evaluation, "that UFO studies do not offer a fruitful field in which to look for major scientific discoveries." Such comments virtually admitted that the analysis given in the report allowed for no explanation of some sightings, and went no further than to state that they could have involved a possible—*but improbable*—confusion with birds, planes or things of that sort. In short, the analysis of the episodes studied ended in an evasive draw, which only left the whole problem intact.

One of the most astounding effects caused by UFOs, according to what witnesses revealed, was the interruption of the electric current feeding a motor car's spark plugs and headlights. Only one such incident was mentioned in the report, which shrugged off the matter with the bland statement that "no satisfactory explanation for such effects, if indeed they occurred, is apparent." This evasion was one Hynek could not swallow. Here was an example of the investigator dodging the issue when it was his mission, after all, to determine whether such incidents had occurred or not; certainly, the Commission had no business to thrust them aside as improbable simply because it could not find any rational explanation for them.

Meanwhile, Hynek showed that the U.S. Air Force kept on its files more than one hundred cases of electric and electro-magnetic interactions between UFOs and motor cars. Yet the Condon report declared that "during the period of field study only one case of automobile malfunction came to our attention." Hynek fully agreed that there was ample reason to doubt the veracity of the case reviewed since it had been reported by a diabetic driving home at three o'clock in the morning from a party at which he had downed many more drinks than he should have. Moreover, the episode had been classified in the first category (priority sightings) when indeed the Commission should have thrown it out of the proceedings.

In agreement with Hynek's avowal that the Commission had deliberately drawn up a wholly unacceptable report, other members took issue with it. Referring to the sighting of a luminous signal, one dissenter declared, "The residue is a most intriguing report that must certainly be classed as an unknown pending further study, which it certainly deserves. It does appear that this sighting defies explanation by conventional means." The Condon group's ranking researcher, Frank Roach, implied that the Commission had left important business unfinished when he declared that "visual phenomena" reported by American astronauts during their space flights "defy analysis."

Among the most impressive cases in the Air Force's files were those in which radar findings coincided with visual observations. Offering no explanation whatever, the Condon report chose to ignore them.

After studying the file of a case according to which a UFO was said to have landed on the ground, Gordon Thayer, investigator for the research services of the environmental sciences, stated, "This must remain as one of the most puzzling radar cases on record and no conclusion is possible at this time. It seems inconceivable that an anomalous propagation (AP) echo would behave in the manner described, even if AP had been likely at the time. In view of the meteorological situation it would seem that AP was rather unlikely. Besides, what is the probability that an AP return would appear only once, and at that time appear to execute a perfect ILS [Instrument Landing System] approach?'' Thayer, in other words, was contending another passage in the report by asserting that the mere possibility of a radar signal being reflected by an atmospheric phenomenon—the standard example of abnormal propagation—was entirely untenable.

Hynek cited another section in the report which practically acknowledged the reality of inexplicable phenomena: ''In conclusion, although conventional or natural explanation cannot be ruled out, the probability of such seems low in this case and the probability that at least one genuine UFO was involved appears to be fairly high.''

To be sure, these few examples were overwhelmingly outnumbered by the multitude of false alarms reported, all of which were easily accounted for. But since they could not be ignored, Hynek protested against the absurdity of the statistics released, intended to liquidate the whole UFO question by emphasizing the enormous imbalance between the relatively few unfathomable sightings and the far greater number of those originating in natural causes. It was not the *mis*identified objects that prompted the investigation, he pointed out, but rather the *un*identified objects.

In Hynek's overall view, UFOs are not simply flying saucers but rather phenomena that require us to expand our knowledge of the Universe around us. This is not the first time the human race has had to face such a challenge. To the man of the pre-scientific era, the Moon, the aurora borealis, the rainbow, tornadoes and light rays were all UFOs.

UFOs Over the United Nations

Where the Robertson and Condon commissions gave the Air Force's UFO documentation little more than a cursory glance, Hynek was the only scientist to examine the complete files, covering some 12,000 cases. But others permitted to scan them also came to suspect that spacecraft of other than earthly origin had indeed been seen in the environs of the globe. Among these favored few were the two men removed from the Condon Commission for their act of delinquency in revealing its secret intentions, and James E. McDonald, at the time Dean of the Institute of Atmospheric Physics at the University of Arizona, who would subsequently distinguish himself for the vigorous campaign he waged in behalf of a fair, impartial probe by qualified civilian scientists.

McDonald, now dead, enjoyed world renown for his contributions to the progress of atmospheric physics. With so many UFO sightings interpreted as atmospheric illusions, he naturally followed the Air Force's UFO communiqués with particular attention. Taking advantage of the Air Technical Intelligence Center's short-lived disposition to open its dossier to competent outsiders, he scrutinized it closely, and eventually concluded that many of the attributions to physical atmospheric causes failed to clarify the circumstances surrounding UFO incidents to his satisfaction.

With the aid of copious notes he had taken, McDonald listed every possible contingency that could explain the sightings reported: 1) fraud; 2) hallucination; 3) erroneous constructions put upon physical phenomenon which are well known to experts but not to others; 4) the effects produced on the atmosphere when space vehicles re-enter it from expeditions in outer space; 5) physical phenomena still only dimly understood even by scientists themselves as, for example, electromagnetic effects, plasma, etc.; 6) psychic phenomena no less dimly understood; 7) extraterrestrial sondes or astroships; and 8) occult manifestations. He then took note of the fact that those specialists who had perused the UFO records over the previous two decades agreed that a substantial number of the reports must be classified in one or another of the first four categories.

But, like Hynek, he saw that once these cases were duly disposed of, there remained an appreciable number of others which had been reported by witnesses whose intelligence and honesty were beyond question. The interpretations given to many of the cases listed in the fifth and sixth categories, both referring to largely unknown physical and psychical phenomena, were poor in logic and lacking in sound scientific foundation. Excluding all recourse to occultism (8), McDonald ultimately deduced that the most likely explanation for sightings that fitted into none of the first six categories could only be found in the seventh—extra-terrestrial origins.

In his opinion, Project Blue Book operated "on a very low level of scientific competence." According to his calculations, the number of unresolved cases counted from five to ten times as many as the Blue Book commission conceded.

In 1965, the foremost private UFO research associations appealed to the United Nations to take action, and one of them urged Secretary-General U Thant to spearhead a movement setting up an international agency which would maintain a constant surveillance of the atmosphere enveloping the entire surface of the globe and analyze the information it gathered. The matter was brought before the United Nations' Committee of Spatial Affairs, to the gratification of McDonald and Hynek, then still consultant to the Air Force, both of whom enthusiastically championed the proposal. The idea was not altogether new to the United Nations, which had considered it, however, from a different viewpoint. Indeed, the General Assembly had already adopted a treaty regulating the exploration and utilization of extra-atmospheric space, in which Article Five hinted at a "preoccupation" with flying saucers by stipulating that all signatory nations must immediately notify all other nations or the Secretary-General of any exceptional phenomena spotted over their territories which could imperil the lives or health of astronauts.

When the Committee for Spatial Affairs set about to prepare for its international space conference in Vienna, the private UFO organizations repeated their petition. Dr. Kurt Waldheim, chairman of the committee, nodded affably, but shelved it.

On June 7th, 1967, McDonald appeared before members of the committee to testify on the scientific aspects of the flying saucer problem. There was good reason, he told them, to lend more serious attention to the "hypothesis that these unusual objects constitute a certain form of extra-terrestrial sondes." Two days previously he had written a letter to U Thant, in which he stated that before undertaking a personal study of the problem, he was little inclined to give credit to such a hypothesis; but after one year of intensive probing, it was the only one acceptable—although still only a hypothesis—in view of the truly astonishing number of low-altitude and short-distance sightings on record throughout the world concerning "objects which have the appearance of machines." To a number of scientists seriously engaged in exploring the UFO phenomenon, the letter continued, it seems most probable that these UFO manoeuvres represent "some type of surveillance probes of extra-terrestrial origin." If there were any possibility, even remote, that this surmise might be correct, then our ignorance of the plans and intentions motivating such a surveillance must give way to an understanding, as complete as possible, of what was going on. If the phenomenon were of some other nature, then we must know it. Our ignorance, our negligence and ridicule, he concluded, are regrettable factors of our collective attitude toward what could well be a matter of urgency for all the peoples of the world.

Testifying before members of the space committee,

McDonald opened his broadsides by citing his own experiences. After a profound examination of the scientific aspects of the problem on the basis of the reports coming from within the borders of the United States; having interrogated the key witnesses in dozens of cases over the years between 1947 and 1967; having observed the Air Force's official procedures of investigation; and having himself verified a great number of other sources of information, McDonald declared, he had come to realize that UFOs were not to be dismissed lightly as so much foolishness but rather viewed as a subject of profound scientific concern, which demanded that "a systematic study should be undertaken without delay on the world scale."

Pursuing his argument, McDonald expanded the theory he had outlined in his letter to U Thant. We are in possession of numerous witness reports, he said, concerning absolutely unknown aerial objects which resemble machines and perform in a way we cannot explain. From what he had learned about them, he could only reject the conviction that they are natural atmospheric or astronomical phenomena which amateurs had interpreted incorrectly. It is no longer reasonable to explain these observations by citing effects produced by industrial technology, secret experimental vehicles, deception or psychological aberrations. Scientists must not cringe before mindless ridicule, but cooperate in a thorough investigation of the utmost precision. Because of the worldwide extent of the phenomenon, it was the United Nations' obligation "to seek all possible means of securing worldwide attention to this problem."

The speaker deplored the adverse reactions engendered by the sarcasm and raillery which distorted the whole UFO question and inhibited a full, impartial enquiry into it. There were too many factors involved that current scientific and technological knowledge were unable to fathom. If these baffling objects were not of extra-terrestrial origin, he warned, then the reality of the situation could be even more bizarre, ominous and urgent. Whatever the ultimate answer might be, nothing less than action on the very highest level could be sanctioned.

McDonald dwelt on the pressing need to employ radar, declaring that no other existing detection technique could compare with it for efficiency in collecting accurate data on the movements and functional characteristics of UFOs. Unfortunately, most of the utilizable equipment, he said, was firmly held in the grip of the military authorities, who refused to admit that UFOs existed; moreover, in most countries, they suppressed all public mention of radar reports concerning flying disks, and most witness reports never reached the scientists.

McDonald reminded his listeners of the variety of electromagnetic perturbations that often occurred when UFOs cruised at low altitudes or hovered motionless in the air, although he conceded that these records could not be accepted as scientific documents. The same perturbations were also affecting internal combustion motors at the very moment low-flying aerial objects passed over them, while radio reception was drowned out by the different kinds of static they caused. In not a few instances, ship and plane compasses were known to go out of order, and magnetometers, clocks and watches affected.

All this evidence suggests some mysterious electromagnetic noise or secondary effect which cannot be identified without the aid of new, much-improved detection devices, but before the competent engineers and physicists can design them, they must study the fast-growing mass of information on short-distance and low-altitude sightings. For the present, however, we must be content with radar as by far the best of all detection instruments.

McDonald was unimpressed by evidence some UFO-ologists had produced to indicate that extra-terrestrials were carrying out reconnaissance and exploratory missions over the Earth; the reports they quoted were too insubstantial, he believed, to support their affirmations. He was no less dubious about the timing of the upsurge in

UFO activities shortly after World War II. No one could say with any assurance whether the UFO phenomenon had become more intensified in 1947 or whether an obsession with it could have made it appear so; but, he said, the frequency of their appearances seemed to have multiplied two or three times since that year "for reasons we do not understand in the least."

Here we must note that McDonald testified not before the official United Nations Committee of Spatial Affairs but rather before its working group, which meant that he held what virtually amounted to a private, unofficial briefing, destined to lead nowhere. Indeed, the committee omitted the issue of UFOs on the agenda of its Vienna conference, no doubt because these space phantoms appear only within the Earth's atmosphere, whereas the purpose of the conference was to establish regulations covering human activities in the extra-atmospheric domain. Moreover, the great powers, as might well be expected, turned a deaf ear to the plea for international control over their national territories. Only a rash or naive soul could ask the United Nations to conduct a program of surveillance and control over the air space of their member nations. One need only recall the impasse that blocked the armaments negotiations when the superpowers refused to permit international patrols to cross their frontiers on what they considered would be snooping missions.

What Happened in the Soviet Union

I have cited only the American Air Force dossier on UFOs because it is by far the most voluminous in existence and because its findings were accessible to scientists, if only for a limited time. But UFO sightings were by no means restricted to the United States; since the war, they have been reported in countries all over the world, including the Soviet Union.

While the cold war was generating deep mutual distrust

between the Western and Communist blocs, Soviet political commentators and journalists complained that warmongering anti-Communists were exploiting the flying saucer issue for all it was worth in their propaganda offensive against the Soviet Union. But their attacks stopped abruptly in 1957, when the national press began to report UFO sightings over Soviet territory. With 1967 a banner year in flying saucer traffic over the country, on 13 October of that year the Government set up a U.S.S.R. Permanent Cosmonautic Commission of investigation, including a membership of eighteen scientists, among them astronomers and one cosmonaut. Air General Porfiry Stolyerov, chairman of the commission, liberally treated his countrymen to a telecast of UFO photographs taken by aviation pilots and private citizens, while the December 1967 issue of *Sputnik* magazine published articles on a number of sightings reported by geologists, astronomers and other scientists.

During the same year, Professor Boris Constantine, Vice-President of the U.S.S.R. Academy of Sciences, sponsored a book on extra-terrestrials, *The Inhabited Cosmos*, one chapter of which bore the title "What Are UFOs?" The author, Dr. Feodor Yuri Zigel, was unable to answer the question, but offeted his readers a measure of comfort by assuring them that the phenomenon was worldwide. He suggested five explanations for them, one of which held that they could be extra-terrestrial spacecraft. A man of caution, however, he made it clear that such a theory, in his opinion at least, lay beyond the limits of common sense. Nevertheless, in the February 1968 issue of the Soviet review *Life*, Dr. Zigel published an essay in which he dismissed all the theories thus far put forth as totally unsatisfactory, and called for the world's scientific community to collaborate on an intensive search for the right one. His summons elicited a hearty endorsement from McDonald.

Official Soviet policy on the matter, however, promptly put Dr. Zigel to shame when the daily newspaper *Pravda* rebutted him with a statement from the Soviet Academy of Sciences denouncing the whole UFO issue as a myth which could not be reconciled with Marxist doctrine. In support of this charge, the Academy cited prior censures by Soviet and non-Soviet experts, particularly an American astrophysicist, Donald Menzel, who declared that eighty percent of all UFO observations were nothing more than deceptive illusions of commonplace events, such as planes in flight, lightning bolts and the like. The remaining twenty percent, he thought, could be dismissed as normal atmospheric conditions seen in exceptional circumstances.

The Academy of Sciences listed all alleged UFO sightings in three classifications: cases of deliberate fabrication; illusions entertained by untutored comrades; and faulty interpretations of natural phenomena. In a follow-up communiqué, it announced that the issue had been fully debated by the members of that august body and found to be nonexistent. The Bureau of General and Astro Physics released a paper by the Academician Lev Artsimovitch, denouncing the flying saucer discussion as sensational anti-Soviet propaganda. All reports, he wrote, were "inventions which have no basis in science; furthermore, the objects seen have all been of a well-known nature." His statement was countersigned by E. Mustel, chairman of the Academy's Astronomy Council, and two other scientists.

Thus, intransigent official opposition to the UFO question prevailed on both sides of the Atlantic, undoubtedly stiffened by the unrealistic demands of scientists agitating for international surveillance of the world's atmospheric space. Now, only private flying saucer societies, hampered by limited means and too often para-scientific, were left to carry on.

Are They Watching Us?

Virtually ever since the hertzian waves were first put to

use for communications, radio astronomers have associated them with extra-terrestrial interventions. Guglielmo Marconi and Nicholas Tesla, both pioneers in the field of radio, believed that they had intercepted messages coming from outer space on these waves. In 1900, the Yugoslav Tesla, a genial electronics expert who first conceived the notion of eventual contacts with galactic civilizations via hertzian waves, with the correspondents both transmitting on the same frequency, announced that he had intercepted intelligent radio-electric signals of an "arithmetical" nature in his Colorado Springs laboratory.

On September 2nd, 1921, the *New York Times* ran a sensational front page story quoting a Marconi collaborator, who had notified the press that the great Italian inventor was reasonably sure he had received messages from Mars on a frequency ten times lower than the one he normally used.

In 1924, Marconi thrilled the world again with an even more startling revelation: he had recaptured a signal amazingly similar to the first message he had radio-transmitted across the Atlantic in 1902, twenty-two years before.

A revived passion for Mars reached its culmination at the time of that planet's closest opposition to the Earth in 1924; every sixteen years or so, that body comes within approximately thirty-three million miles of the Earth, when observation is most propitious. At the suggestion of an astronomer, David Todd, the United States Chief of Naval Operations ordered the twenty most powerful radio stations on the Navy's American bases to suspend all routine air communications and listen for any unusual signals that could be attributed to Martian origin. The Army took similar measures, and held its coding sections in readiness to decipher the hoped-for messages. About one-half of America's commercial radio stations joined in the experiment, and agreed to go off the air for five minutes every hour during the period of Mars' closest opposition.

During the experiment, some wireless operators persuaded themselves thay they were indeed tuned in to Mars, but to their dismay, the signals they picked up were easily identified as purely natural. Mars, in fact, maintained a total silence; and the American people, deluded by that planet's stubborn lack of cooperation, turned their attention to other diversions. Still, a number of wireless operators refused to believe that the signals caught by Marconi and Tesla were not bona fide replies to the transmissions the two men had beamed to Mars; moreover, they were deeply impressed by Marconi's conviction that fragments of signals he had intercepted from another planet repeated the historic message he had sent across the sea so many years before. It was indeed a sore temptation to accept the word that extra-terrestrials, whether Martian or other, were relaying his own signals back to Earth.

Eager men everywhere looked for this mysterious, thrilling event to recur elsewhere, perhaps in different guises. During a test conducted in 1902, men aboard a ship on the Potomac conversed by radio telephone with a group of Congressmen on the river bank half a mile away. In 1924, at the time of Mars' closest opposition, Todd, with the aid of an optical recorder he liked to call a radio camera, intercepted radio-electric signals that simulated vague figures. The more quixotic wireless operators needed no encouragement to interpret this as a cycle used by extra-terrestrials to return historic transmissions after a precise lapse of twenty-two years.

Today we might ascribe such naive concepts to an engaging folklore and the gullibility of some scientists and technologists. But more serious was an article published in a 1928 issue of the sober, respectable British scientific magazine *Nature*, written jointly by Carl Störmer, a Norwegian mathematician, and a Netherlands specialist in telecommunications, Balthus van der Pol, who were experimenting at the time with radio soundings of the ionosphere. In their essay, these two researchers told of having repeatedly intercepted the echoes of their own signals after several seconds or minutes.

Similar cases reported by other specialists led to the belief that their radio transmissions into the void were being received somewhere in outer space and relayed back to them.

No explanation has ever been found to clear up this riddle, although one theory suggested that layers of ionized gases collected somewhere in the solar system, being reflective, caused the radio waves to rebound; few, however, were disposed to accept this notion. Others clung to the conjecture that the delayed echoes could only be explained by the action of highly alert radio operators in outer space, who were retransmitting signals they had intercepted.

Caught up in this fever of hypotheses, in 1974, a Scottish astronomer, M. Duncan Lunan, dug up the Störmer-van der Pol report and mulled over it to see if he could read some hidden meaning into the signals it described. He succeeded beyond all his dreams—or so he thought—when he traced a graph with two sets of numbers indicating the radio-electric impulses recorded by the two authors and the delay time of the echoes. The result of this transposition overwhelmed him: the designs that emerged clearly represented the constellations in the Northern hemisphere. Especially recognizable were the Big Bear and part of Boötes. Since the anonymous cartographers had indubitably mapped our firmament, Lunan figures, they must have been operating somewhere in the vicinity of the Earth.

The Scotsman was even more astounded when he discovered that each map defined the configurations of the stars such as they must have been thirteen thousand years ago. This inference would seem altogether reasonable if we consider the deformation of the constellations caused by the deplacement of the stars in the heavens.

Lunan scanned other documents on delayed radio echoes confirming his version, particularly a series of reports dating back to 1923 prepared by a French expedition in Asia to experiment with the propagation of hertzian waves. In the graph he composed from these data, he found a message which he translated as: "Our stellar system is Epsilon of Boötes, a double star. The bigger star has seven planets. We inhabit the sixth, which has two moons. The fourth has three moons, the first and third one each. Our sonde is in orbit around your Moon."

Lunan knew that Epsilon of Boötes, 103 light years away from us, no longer belongs to the "principal series" of stars, as does the Sun.[1] Therefore he inferred that the population of that aging planet had sent out a sonde to shop around for another more accommodating habitat.

A team of scientists set out to track down the sonde that Lunan suspected of surveying the Earth while orbiting around the Moon. To seek it out, they turned their attention to the disturbing issue of delayed radio echoes. Under the direction of Anthony Lawton, expert in communications, the researchers employed a radio transmitter sending pulses into various sectors of the sky; simultaneously, a receiver placed in a different locality tuned in on the same sectors. All the echoes were then recorded and compared. The results showed that the delays, ranging from one to twenty-eight seconds, were caused by a tension in the ionosphere. This tension, in turn, was created by an exceptional alignment of the magnetic fields surrounding both the Earth and the Sun. The phenomenon occurred most frequently during the months of November, December and January.

The outcome of the team's efforts proved to be negative; if any humanoids were looking us over from their lunar perch, it missed them.

The experiment thus shattered an ingenious theory, which Lunan had pumped up and hoped to make credible by his "discovery" of a relatively complex message buried in a diagram of inexplicable echoes. But then, one can read all kinds of imaginary messages into any series of dots distributed one way or another over a "graph"

[1]The "principal series" of stars includes the greatest number of suns, all relatively young. Epsilon of Bouvier, in decline, is no longer one of them. Any life it might support cannot endure for very long, hence its inhabitants would be in search of new quarters.

A Technological Religious Myth

Scientists have not yet found any solid evidence that extra-terrestrials are piloting their spaceships in the proximity of the Earth to have a look around; and those individuals who continue to believe that outer-spacemen are dissecting us through their futuristic spyglasses tend to accuse NASA gratuitously of withholding urgent information and imposing a total blackout on momentous discoveries made by its scientists and astronauts. The latter, according to subterranean rumor, saw far more in their excursions into space than the American authorities are willing to tell the world. But aside from NASA's alleged secret files, there is another very likely explanation for the missing disclosures: like hunters stalking their prey, our visitors from outer space are deliberately keeping themselves well out of sight.

The assumption that they are circling our globe has gained ground in the public mind, and the fact that no one has ever been able to confirm their presence beyond any possible doubt is attributed in some quarters to the extreme lengths to which they go to avoid detection. But how can we explain that these cautious spies do occasionally dare to give some slight hint that they are indeed hovering above us? The answers to this question are highly contradictory.

Thus, the whole question of snooping extra-terrestrials remains, for the present, a matter of personal intuition.

The U.S. Air Force has been sharply criticized for the undue zeal of its psychologists—there were far too many of them, some say—in grilling the good folk who confessed to having spied a flying object over a hill or a meadow. Were these witnesses motivated by wayward of hyperactive intuitions? Were the stern psychologists justified in maintaining their inflexible attitude of skepticism? Do the few scientists engaged in the study of extraterrestrials hold to a balanced objectivity in their approach to the subject? Can we take what they tell us seriously or do they arrive at their conclusions under the persuasive but misleading dictates of their personal emotions?

Admittedly, a researcher's task is more formidable and complicated than many of us suspect. He must be vigorous and alert in exercising both his critical sense and his imagination, pitting his skepticism against his intuition. A hypothesis grows out of a preconceived notion that germinates in a terrain of dissatisfaction with prevailing habits of thought and standards. Unfortunately, in the case of extra-terrestrials, there are no standards to go by, despite the valiant efforts of UFO investigators everywhere to establish them.

This, however, does not discourage some enthusiasts from formulating conjectures rooted in their personal emotions alienated from scientific fact. To them, extra-terrestrials have become familiar demons indispensable to some inner need, or a narcotic addiction, an *idée fixe*, whatever the extent and depth of the intellectual resources he brings to his contemplation of the question. The great psychoanalyst Carl Jung was deeply intrigued by the passions aroused by flying saucers, and powerfully attracted by the collective aspect of this psychic manifestation.

Jung, however, never openly declared himself on the practical aspects of the problem or the facts. In his view, "the physical reality of flying saucers has not been determined one way or the other". The nub of the question was something else: what interpretation must be put on the "immense visionary rumor sparked by this uncommon emotion?" A whole myth revolving around extra-terrestrials, he observed, was in the process of evolving. Orson Welles' much discussed pre-war radio dramatization of H.G. Wells' *The War of the Worlds*, narrating a fictional Martian invasion of the Earth and the terror it spread, offers an illuminating example. The imminent threat of another global conflict had exacerbated the anxieties of his audience. In the oppressive climate of the ensuing cold war, the mysterious flying objects that suddenly appeared in the sky were associated with the threat of an invasion by hostile outer-spacemen. But gradually the preference of our invisible guests for total anonymity and their docility—rare were the episodes of violence reported by observers—combined to lend a new profile to the situation, and people everywhere began to believe in extra-terrestrials out of a *will* to be surveyed, to nurture the feeling that the cosmos was somehow watching over them.

Jung stated that the emotional tension attached to these "rumors" arose from two sources: a situation of collective distress and a vital need of the soul.

Under the pressure of the cold war, Jung wrote, "the condition of the present-day world, in which the individual begins to feel and understand that anything and everything can come into question, is so pregnant with dire threats that the unconscious imagination, which invents projects, rises above terrestrial powers and organization to soar into the sky, into cosmic space where, in other times, the gods, masters of men's destinies, maintained their headquarters, somewhere among the planets."

In other words, the world's apocalyptic atmosphere was reviving the function of religion in a society whose religious institutions had become inadequate, having lost their capacity to satisfy the profoundly human need for the solace of mysticism.

In modern times, what is sacred must be sanctioned by science. The marvellous finds an easy explanation in the recent triumphs of technology; what has always been thought forever beyond man's reach has already been achieved. Flying in the air, television, computer brains and other miracles have become commonplace events that no longer stir anyone's wonder. It is now permissible to acknowledge the feasibility of the impossible and to explain it. A veritable legend has grown up around flying saucers—"a living myth," according to Jung.

This "living myth" encourages us to welcome our friends from other worlds just when we are beginning to dream about escaping from our own and penetrate the Universe. To Jung this two-fold aspiration was significant of the world's current state of rupture. Those thrusts of the imagination, the motivating impulses which influence our thoughts and, oftentimes, our actions—a hunger for sensation, a passion for adventure, intellectual curiosity, the ecstasy men find in miraculous technical performances—are almost always the consequences of a vital distress situation. This distress, intensified by a growing awareness that our globe is too small for comfort, translates into a dream of evasion. In an inimical world, anguished human beings turn to extra-terrestrials for relief. "This is why", Jung points out, "the signs of heaven appear" in a form that fulfills the demands of a scientific fantasy.

Our explicative "projections," therefore, are born out of an anxiety which, anchored in motivations only dimly perceived, remain in the subconscious. Without doubt, "collective rumors" have existed in every era of history, but to find the mainsprings of a phenomenon as bizarre as the flying saucer myth, "we must dig down deep to the very roots of our existence."

Without attempting to offer any sort of scientific interpretation to the air-borne "round bodies," which Jung readily conceded were "sometimes" registered in blips on radar screens, the eminent Swiss psychologist saw them as symbols of perfection, of the absolute: a sign, an image of God. The essence of Jung's exegesis is that, adapted to the modern creed, God—or any absolute or promise of ultimate salvation—assumes the aspect of a *thing*, or a technical form. A man will superficially identify God's powers with those which are typically man's as if "to avoid the indecent revolt of a mythological personification." In Jung's view, modern man rejects the concept of a personalized God, such as the God of Israel. He replaces Him with a technical symbol associated with His omnipotence.

The contemporary climate of the world, Jung believed, is such that none other would be as likely to arouse universal hope and expectations of some supra-terrestrial event that would submerge our latent conflicts. A spaceship from another world is the mediator, the surrogate for a Christ no longer recognized as the Messiah, the latest in the ranks of those traditional messengers who, like its predecessors, will assure mankind that the bonds between the gods and men remain safely secure.

THE ODDS ON LIFE IN SPACE

The Viking Program

Planetology, heretofore largely a matter of difficult long-distance observations and hypotheses, has been entrusted to automatic spacecraft launched on exploits that yield far more accurate information. These electronic wonders orbit the planets under study, photograph them at close range and land on their surface to scoop up soil. American Apollo expeditions "in the field" have bagged hundreds of pounds of such samples. Soviet sondes meanwhile have managed to keep their transmitters functioning for well over ten minutes after alighting on the deadly surface of Venus, relaying first-hand data that no laboratory work could match for precision.

For the first fifteen years of space soundings, such investigations were confined to the three celestial bodies nearest to us, the Moon, Mars and Venus, but NASA probes have begun setting their sights on the huge planets farther away, as well as on little Mercury. In a new economical development, space engines are programmed to perform multiple services by delving into the secrets of more than one astral body on each excursion. Mariner X flew past Venus and photographed it before going on to make three successful orbits of Mercury. After Pioneer X's historic flybys of Jupiter, Pioneer XI took the same route to that giant planet, then veered off to Saturn, which it will orbit in 1979. It will not depart the solar system like its predecessor but, powered by Saturn's energy, escape the orbital track of the planets and go on to explore a region in space of which we know nothing, above and below our rotatory course. Less spectacular but nonetheless meritorious are the space missions sent out to uncover important information on magnetic fields, solar winds and various other natural phenomena.

With this invaluable assistance of its sondes, NASA is now prepared to explore the whole of our solar world. Leading the procession to the outer planets, the two Pioneers will be followed in 1978 and 1979 by the other more sophisticated craft of the Mariner type. New Pioneers will set up laboratories on the surface of Venus to scrutinize its soil. In the 1980s, more ambitious sondes will reconnoitre its atmospheric and surface characteristics in greater detail. For their part, the Soviets are apparently preparing to explore other aspects of the planet not included in NASA's program.

One of the main objectives of these flights is to determine whether living organisms—or at least the chemical conditions necessary to support life—might have evolved on any of these globes. The question has been titillating many an exobiologist ever since a series of Miller-type experiments established that any simulation of primitive atmospheric conditions focused on substances thrown off by the original solar cloud will bring forth molecules of biological value.

Several years ago, the leading Russian and American scientific institutes agreed that research into extraterrestrial life rated top priority in their unmanned space programs, but that spatial techniques had to be perfected on the interplanetary scale to meet the challenge. The exobiologists profited by these years of apprenticeship to prepare the way for a comprehensive program of detection. The Ames Research Center, engaged in the study of outer space life forms, joined forces with specialized university centers to organize a series of convergent projects. With some biologists examining carbon compounds taken from meteorites that have landed on Earth, others have been sifting through data collected by space probes into the atmosphere and ecology of the planets surveyed by sondes, from which to build models of them and reconstruct the probable course of their evolution. They have also tried acclimatizing terrestrial organisms to simulated outer space environments, notably that of Mars, some adapting to them with surprising facility.

Since the Martian ecology most closely resembles our own, it attracted the major share of attention; and scientists wondered whether the planet's chemistry had progressed to the point where it could germinate living organisms or whether there were any traces of chemical conditions that tend to structure the macro-molecules which create living things on Earth. To approach the problem, it would be necessary to observe certain chemical transitions inherent in life forms, to find indications, for example, of the basic mechanisms by which the living organism synthesizes or acquires energy molecules, which it breaks down to obtain the energy or spare parts it needs.

Getting this Mars project figuratively off the ground, in a preliminary phase, teams of biochemists, astrophysicists and cyberneticists drew on all the data available on the solar world. Helped by new analytical methods, they studied the lunar rocks brought back by the Apollo astronauts.

Contrary to some biologists' expectations, it was found that the carbon molecules on the Moon's terrain never passed beyond the first stages of synthesis; before they could produce amino acids, their evolution was thwarted by lack of water. Furthermore, it is probable that these rudimentary precursors of life were not compounded on the Moon at all but borne to it by solar winds. In any case, an insight into lunar chemistry will not help much in predicting what the sondes will find on our sister planets.

With a billion-dollar investment in its Viking program directed to Mars, NASA cautiously planned a dual launching to increase its margin of success. Two similar sondes, Viking I and Viking II, were scheduled to make soft landings on that planet in the summer of 1976, each on a different site, over five thousand meters below the planet's median level. If water in liquid form existed in the depths of these plunging abysses, the best place to look for it, then Martian life might possibly exist in some form.

The first site is located at one extremity of a vast fault, thought to be an ancient course into which most of the water could have drained from the orb's equatorial regions. The second site holds out more promise; still deeper, it lies at one end of a mist blanket, perhaps formed out of water crystals, that envelops the polar region during the Martian winter.

In their priority mission—to search for vestiges of past or present life, or perhaps chemical structures arrested in their escalation toward living organisms—both vehicles will orbit the planet, functioning as observatories and relays, and later eject a one-ton robot lander to descend to the surface. Mounted on three legs, both landers will be equipped with two color cameras and a mechanical arm three meters long to scoop up the soil for on-the-spot analysis in a fully automated laboratory.

The Soviets undoubtedly have another project in mind: to send up a Lunokhod type of sonde, charged with bringing back samples of the Martian soil, in an elaboration of the techniques they used in their Earth-Moon soundings carried out by Lunik, Luna and Sond automatic craft.

Needless to say, the far longer distances—hundreds of millions of miles rather than the comparatively short hop to the Moon—will pose a good many more difficulties. On Martian soil, the robots must command a considerable degree of autonomy to determine for themselves what actions to take, depending upon the information they collect. Thus, space technology has moved on from remote control of its vessels to highly cyberneticized self-regulation. The ground crews will do no more than plan the programs and trigger the sequences in good time when necessary.

The Riddle of Life on Mars

Our nodding acquaintance with Mars is ripening into a more intimate familiarity. In 1965, the world was staggered by Mariner IV photographs revealing a stark, desert globe of scattered craters, pitted by bottomless canyons, much like the lunar landscape, hardly a congenial habitat for little green men with antennae sprouting from their heads. Two years later, Mariners VI and VII not only disproved the myth of Martian canals but also entrenched notions about the planet's atmosphere. With no discernible traces of nitrogen, their soundings showed its essential elements to be carbon gases with a dash of carbon monoxide, and its density, previously overestimated, somewhere between 1/50th and 1/150th that of the Earth at sea level, and containing very little water. Surrounding the planet is a 12,000-mile crown of hydrogen, perhaps resulting from the disassociation of water vapor molecules drifting off into space. If this has been going on over a period of eons, then Mars must have been once covered by vast oceans. From additional data signalled by Mariner IX, experts believe that permanent sources of water probably still exist.

Another significant discovery showed that Mars' magnetic field is no stronger than the Moon's.

The planet's relief is very volcanic, and past eruptions may have had a good deal to do with the nature of the atmosphere. Some volcanoes could still be active. Those on Earth spew up steam and carbonic acid, substances essential to life; and lava is a fertile growth medium. Therefore, Mars' history of spectacular volcanic upheavals offers a shred of evidence that some form of life may have evolved.

The NASA sondes yielded other important data. Mariner IX measured surface temperatures from –130° F at the north pole to 77° F at the equator, indicating that ice can melt in some areas. The low-lying polar regions apparently retain their atmosphere better than elsewhere, and air pressure measurements confirmed by the craft show it to be denser there. But it is possible that geological accidents in other areas may have caused surface cave-ins, creating zones where the air is less rarefied.

The Mariner IX photographs of the planet's immense canyons enabled scientists to identify the network of tributaries that once fed these water courses. There is probably ice on the surface, but the polar caps are held to be the principal sources of water, combining carbon-dioxide snow, ice crystals and liquid water molecules. The northern cap is unstable, and the tilt of Mars' rotation, together with the erratic course of its orbit, varying between 125 and 150 million miles, accentuates the differences in temperature between the two poles. During the sonde's visit to Mars in its springtime season, water was suspended over the austral cap but not over the northern cap.

Sagan has produced evidence that Mars once enjoyed a climate like ours, with flowing water, higher temperatures and a denser atmosphere than it has today. But water can no longer be retained because it would evaporate in the thin atmosphere. In ages past, however, when the climate was warmer and the air heavier, life forms similar to some on Earth could have existed.

Scientists now believe that pre-biological compounds appeared in the empty space which later become the home of our solar system. On the young planet Mars, these compounds might have combined with organic substances synthesized under favorable environmental conditions. Did this process come to a dead stop before it could produce living organisms? If not, what forms did life take? Experts agree that any existing Martian vegetation could have descended from more numerous ancient species, most of which disappeared as the ecological conditions worsened. Did Martian vegetation follow an evolutionary pattern parallel to ours?

On Earth, all living things apparently grew out of the same root, and with the identical basic molecular structures which, with time, developed into a multiplicity of forms. The biogeneticists attribute this diversity to a long series of evolutionary constraints and adaptations brought about by genetic accidents that altered organic cellular patterns. The exobiologists gave special attention to these environmental constraints.

Laboratory experiments to test the adaptive faculty of living things in a simulated medium have proved that many micro-organisms can survive in a freezing, arid Martian climate lacking oxygen in the atmosphere. More water was all they needed to grow and multiply.

Similar living things on Earth have had to endure conditions of the same sort, but they were—and are—geared to cope with them. Many bacteria develop in airless media, like the microbes responsible for botulism that proliferate in sealed jam jars. But this is not to say that such bacteria, which feed on organic matter, exist on the red planet. Earthly life has many other formulas, microbes, for example, that subsist on inorganic matter like iron, sulphur and hydrogen sulphide, plus a little oxygen from the atmosphere. Terrestrial life forms may also include micro-organisms that depend upon nitrate as their oxydizing agent; but is there nitrate on Mars?

If there is water, however, then we need not refer to primitive terrestrial organisms; plant forms taking carbonic anhydride from the air can easily keep going. Rye, barley and groundnuts have survived a Martian climate. Some forms have even altered their metabolic mechanisms to produce carbon monoxide and hydrogen instead of manufacturing oxygen in the normal way.

Far more complex creatures have been known to resist formidably hostile ecological conditions. Tortoises have survived almost two months under atmospheric pressure equivalent to an altitude of twelve miles, which would be fatal to a man. The ostensibly defunct larvae of the scarab beetle revived when they were restored to normal conditions.

Of all living organisms, lichens would best adapt to the Martian world. Growing on rocky mountain slopes with almost no humidity, they can survive immersion in liquid air up to –320° F. Some species live comfortably in permanently gelid areas, while others thrive in the Sahara, where no rain relieves the blistering heat. Some lichens erode the rocks they cling to by extracting the minerals, and other species draw nitrogen from the atmosphere. On Mars, these organisms could exploit the humidity suspended in the atmosphere in the form of minute crystals, or absorb the water in the air despite the very low level of steam.

Most significantly, certain lichens can stock oxygen and liberate it through photosynthesis within their porous structure. This they do in high altitudes to boost their oxygen reserves. A plant that fabricates its own carbohydrates by photosynthesis in the light needs oxygen to release the energy stored in its metabolic processes. Possibly lichen-like plants on Mars accumulate the oxygen they have produced instead of freeing it, and draw on it only when necessary to liberate stored energy or to activate replacement matter. The intense cold could benefit these organisms by slowing down their metabolic processes to ensure survival in an environment where food is scarce. Such hibernation would help them hold on to their oxygen reserves.

Are these hypotheses mere shots in the dark or valid, at least in part? NASA's Viking program will enlighten us on some, if not all, aspects of the Martian riddle.

The Impossible Planets

Although Mars is probably the only other planet in our system on which life may have taken a foothold—

therefore representing a priority field of investigation—some exobiologists are searching for traces of it even on the most hostile worlds.

Taking Jupiter as an example, Dr. Cyril Ponnamperuna, one of NASA's ace exobiologists, theorizes that it may have accumulated prebiotic compounds, at least in the upper layers of its gases. This giant orb has no crust; its surface is awash in a vast ocean of liquid hydrogen, with a composite atmosphere overhead of primitive gases from its original cloud: eighty-two percent hydrogen and seventeen percent helium, water, ammonia and methane—a world in sharp contrast to the Earth. (Astrophysicists believe that Jupiter is a would-be star that failed to take fire—that is, it never set off thermonuclear combustion—because it lacks the requisite mass.) Ponnamperuna suggests that some pre-biological processes may have evolved in its atmosphere. By reconstructing this supposed atmosphere and simulating the electrical discharges of Jovian storms, he and other scientists have reproduced compounds of a highly important biological nature, including prussic acid. But such hypotheses are extremely tenuous, and it is difficult to see how life could have taken root in these frozen globes so far from the Sun.

Some exobiologists and astrophysicists, however, contend that life need not be excluded on planets much different from our own, and Sagan counsels us against "terrestrial chauvinism." Why should non-telluric planets be arbitrarily dismissed as incapable of supporting any form of life? They are cold only on the surface; below the outer layers of gas, we find a rising temperature. Observing the Earth through an infrared telescope equipped to measure its heat, an outer-spaceman would most likely conclude that it, too, is a cold planet. At a certain depth in the atmospheres of Jupiter, Saturn, Uranus and Neptune, he thinks, it is possible to find areas with temperatures "comfortable" by our standards, and perhaps another kind of biochemical organism which has managed to defy climates and pressures very different from ours.

Let us now consider the telluric planets, those small, hard-crusted planets close to the Sun and akin to the Earth. Oxygen is their number one component, hence with their hydrogen, water should be their most prevalent compound. Indeed, these planets once produced quantities of water, released in abundance by their rocks; some have retained it, but others have lost it through evaporation in scalding temperatures and weak gravity. Their atmospheres, in effect, dispersed in space.

Mariner X has told us something about Mercury, nearest to the Sun, where ground temperatures soar as high as 700° F. Too hot and feeble in gravitation to retain water, this arid, burning ball has expelled most of its gases. Life here would seem unthinkable, yet there are exobiologists who say that some sixteen feet below the surface, the temperature could drop as low as 90° F, even in the equatorial belt. Theoretically, water could exist at that depth, or perhaps hydrous rocks that make life possible.

Although farther away from the Sun, Venus is nevertheless more torrid. Soviet sondes landing on her soil recorded a temperature of more than 860° F. This heat results from the heavy cloud cover of the dense atmosphere, which veils its surface from our view. Water in liquid form, which could not exist either on or below the surface, is found, however, in high altitudes. The Venusian atmosphere, including its clouds, consists to a great extent of sulphuric acid. At ground level, it is an infernally hot, corrosive smog, almost visually impenetrable. Still Sagan thinks we cannot exclude life entirely on Venus; here on Earth, he points out, there are bacteria that feed on metal rust, sulphur and methane. Perhaps, he says with a wink, the Venusians are a race of demoniacal winged creatures, protected by a hard-shell coat, who fly through the thick air.

Let us avoid undue pride in our terrestrial superiority, says Sagan, and allow that, where life cannot be definitely discounted, it could well exist in forms totally alien to those on Earth. It is possible that life has conformed to different circumstances or developed means of survival at variance with ours. While alternatives to carbon, like silicon and germanium, which share its fundamental properties, have been proposed as the basic element in organic chemistry, carbon can combine with a far greater number of potential compounds. It is also more common in the cosmos than any other candidate element. Experiments in synthesis reconstructing primitive atmospheric conditions whether on Earth, on Mars or Jupiter, show that nature prefers carbon, from which it spontaneously manufactures amino acids.

It would seem, then, that extra-terrestrial organisms consist of the same molecules as those on Earth, although life may have followed different evolutionary processes to befit environmental exigencies. Scientists are now inclined to accept the principle that all living things have come out of the same melting pot, requiring only a few fundamental patterns of evolution; and they agree on the astounding adaptive qualities of living organisms rather than polymorphic origins of life.

What Radio Astronomy Found in Space

Through the medium of radio antennae, unprecedented scientific strategy has brought forth a startling revelation —that a whole chemistry similar to our own also thrives in open space, that space is no longer that arid, inert void blocking all chemical exchanges, as everyone believed it to be, and that its low temperatures are no barrier to exchanges among simple bodies. In the interstellar expanses where interactions are infrequent because of the rarefaction of gas, nevertheless atoms and molecules do appear in very different states of energy which render them reactive despite the frigid climate. Hence, temperature is no longer an index of chemical activity; what counts is the excitation level of the atoms and molecules.

Optical astronomy had already begun to discover that,

like the Sun and the stars, interstellar space contains a chemistry of atoms and molecules in the form of rarefied gases. Since 1937, the astronomers at Mount Wilson have been finding molecules consisting of two carbon atoms combined either with hydrogen or nitrogen. In 1951, American radio astronomers discovered the emission of cold hydrogen on the twenty-one-centimeter wave length. Indeed, a probe in depth of this fundamental gas which structures our entire Universe made it possible to map the first chart of the Galaxy.

Thenceforth, radio telescopes relayed optical observations for the chemical analysis of bodies suspended in galactic space. Spectroscopid analysis (of the spectrum of light rays caught by optical telescopes) has its counterpart in the radio-electric domain; atoms and molecules emit on wave lengths that define their "signature" and enable us to find them.

In 1963, radio astronomers detected the radio-electric waves of hydroxyl—an atom of oxygen and another of hydrogen—which is nearly water.

Water itself was discovered in 1968 with ammonia—the first molecules of three and four atoms to be found in space. With this momentous development, the panorama now began to widen rapidly. Two dozen more complex molecules were subsequently observed, some of them comprising as many as seven atoms. Among these bodies were counted numerous molecules that belong to the chemistry of life, formaldehyde, methylic alcohol and monomethylic acetylene, to name three. Such molecules, which usually turn up on very short millimetrical or centimetrical waves, were probably formed on the surface of spatial dust, from which they eventually detached themselves.

Today radio astronomers are scouring the firmament in the search for units more complex than these "organic" molecules, but the exobiologists' laboratory experiments leave little room for hope; apparently a solvent environment and a liquid phase are indispensable to a process before it can produce richer structures.

Another important source of information on this subject is obligingly brought to us by meteorites. These natural objects falling on our soil come from the asteroidal zone, where thousands of fragments of a shattered planet, once in orbit between Mars and Jupiter, are still circulating. Unfortunately, it is difficult to tell whether the specimens found on these celestial rocks and analyzed actually come from outer space, the pollution of the atmosphere or the soil, or whether they were polluted in handling. Meanwhile, it appears that Ponamperuma, one of NASA's top exobiologists, has proven the extra-terrestrial origin of amino acids borne to Earth by a meteorite.

Thus, since these fragments were once integral parts of a vanished planet, the amino acids they carry could have come into being on that planet during a secondary formative phase of the liquid environment.

The discovery in space of chemical bodies already structured gives further valid support to the theory that the escalation of chemistry toward living matter follows a single—and necessary—route. To be sure, these are only relatively simple structures composed of a few atoms, far removed from the giant molecules—the proteins and amino acids—that constitute living matter. But it is clear that a carbon chemistry identical to that which chose life on Earth, however rudimentary its forms, exists in rarefied open space; therefore it is logical to assume that more complex processes climbing the ladder toward life will have evolved in the dense atmosphere of planets in other solar worlds. The chemical information collected by our radio antennae thus appreciably endorses the arguments based on probabilistic calculations.

The discovery of carbon chemistry in space, moreover, would suggest that any extra-terrestrial life must have narrower chemical bases than we should have hitherto thought likely. Today, the exobiologists are more than ever inclined to accept the postulate that our terrestrial biochemical model is universal.

Do Galactic Societies Endure Long Enough to Signal Us?

Ever since the first powwows at Green Bank and Byurakam, American and Soviet astrophysicists, constantly studying and restudying possible models of advanced civilizations, have been pondering a logical question: aside from radio signals extra-terrestrials might be sending out to other extra-terrestrials (also, perhaps, to terrestrials), can our observation instruments detect traces of their purely local planetary pursuits in residual vestiges of their industrial and electronic activities, and the like? Here again, the only way they can approach the problem is by reference to the Earth model.

In reverse, any imaginary humanoids who might be eavesdropping on us from some other planet in our solar system could not fail to take due note of our radio emissions, even those not particularly sent in their direction, and the movements of our space vehicles. Indeed, these automatic emissaries have given us ample proof that no such onlookers exist, but perhaps there are others who, from their far-distant home bases, are watching the rapid transformations man is making on his planet and in his environment.

In accord with Kardashev's assumption that technological civilizations, sooner or later need the total energy produced by the solar systems they occupy, the physicist Freeman Dyson tried to figure out how our society, having arrived at this stage, could resolve the problem. The logical solution, he thought, would be to harness all the Sun's radiations by enveloping it in a huge construction built of matter taken from some enormous but useless planet—Jupiter, say—and shaped in the form of an egg shell. A part of the energy diffused by the Sun and utilized by our emigrant earthlings dwelling in the new biosphere they will have colonized, must eventually be irradiated outside the sphere at a temperature of 300° K, corresponding to a wave length of ten microns in the radio-electric spectrum.

This is why Dyson counsels astronomers to accept the infrared stars as the possible hosts of advanced civilizations, and suggests that besides attempting to establish radio contacts with them, another effective way to seek them out would be to launch a comprehensive research program into the precise origin of infrared radiations.

If half of all galactic societies turned out to be well ahead of us in the business of exploiting the energy of their solar systems, then we should expect to see some indications of the uses they make of it. The energy of a system cannot be depleted; it can only be converted from one form into another, from a higher to a lower level, and ultimately expelled into heat. This unused energy should be detectable.

Is this technical escalation possible for galactic civilizations? Such a fantastic hypothesis is by no means gratuitous, by Dyson's estimate. With a rapid propulsion into the technological era—a very brief evolution in terms of geological time—this mighty engineering feat, he avows, can be within our grasp sooner than we think.

The requirements for an audacious undertaking of such proportions are appalling to contemplate. Yet within one thousand years, he warns the world's population growth, even underestimated, and its correlative demands for energy will render life on this planet altogether intolerable. Within the coming millennium, man will have exploited the total energy provided by the Sun and the zones that favor life.

Mature societies that keep their dynamism going and take every precaution to avert atrophy and mass suicide must, of necessity, emigrate from their solar habitats to platforms in other systems where they will find the energy they need. Dyson agrees with Kardashev: for those societies whose planets can no longer sustain them only a sidereal expansion, first on the stellar, subsequently on the galactic scale, will enable them to survive.

Why then, von Hoerner wonders, have our radio astronomers never sighted any evidence of it through our most powerful telescopes? Could it be because life and intelligence are not, after all, as extensive throughout the Galaxy as we supposed?

An affirmative reply will give one explanation. But there are others. Perhaps these societies have chosen to abandon technological growth in favor of altogether different goals (perhaps taking this decision before their scientists could get around to interstellar contacts). Then, too, a technical civilization is vulnerable to bitter internal strife, which can plunge the entire population into a grave crisis, and even bring it to a state of collapse and its ultimate demise. To survive a global disaster of such magnitude, it might voluntarily curb its technical development and lapse into a kind of intellectual Malthusianism. An overpowering will for peace and tranquillity could compel it to call off its calendar of scientific explorations, such as space programs, and embrace a utilitarian existence that excludes any further adventures.

The stabilization of science and technics at a fixed level and their assignment to strictly mundane objectives derive from the population's loss of interest in continuing progress or from a collective desire to inaugurate a new way of life that rules it out. This state of affairs, von Hoerner believes, could be precipitated by a series of socio-economic failures, a demographic explosion, biological degeneration, a trend toward mass self-destruction or a policy of stagnation wilfully adopted as a "final solution" to harrowing problems provoked by technology. For the communications civilizations, he perceives only a thin, dividing probability line between zero development and swift self-destruction, and believes that all high-powered galactic communities—and our own—live under the dark shadow of these contingencies.

In such a gloomy context, telecommunications become a vital factor of hope and self-preservation—to the German-born mathematician, a mighty challenge of the mind substituting for physical expansion into space and the conflicts and self-destruction it engenders. Interstellar contacts imply no threats, whereas intergalactic travel, if such a thing is possible, will kindle an obsession to annex living space and lead to the explosive consequences of colonization. The former demands a high degree of technological dynamism and invention, at the opposite pole of stagnation and intellectual disengagement.

Von Hoerner deems current hypothesis on the probable density of galactic communications traffic extremely uncertain, especially because it is so difficult to arrive at any valid conclusions on how long our interlocutors can take to establish connections. One discouraging theory holds that a permanent galactic network most probably does not exist, and that other communities are still at our same initial stage of experimentation. Should this be true, then our radio astronomers would have to tune in separately on each of the candidate stars within a radius of one thousand light years.

But the optimists take issue and put their faith in well-organized networks. Intelligent societies, they retort, have been developing mutual contacts for millions of years; consequently, it should be possible to capture signals beamed to us—or traces of local activities—with a radius of one hundred light years.

What then, are the chances for success in our campaign to join up with our faceless friends in outer space? The future, perhaps, will tell us. Meanwhile, there are various considerations to ponder. How many societies? For every one hundred suns in our Galaxy, it is thought likely that at least one will have a planet in tow with conditions favorable to the higher forms of life; but the number on which civilizations on a par with ours can have actually evolved will certainly be far fewer.

How long do they exist? Inclining toward the Soviet view, von Hoerner shares their belief that a technological community can endure for several hundred thousand years before it becomes extinct. Estimating on this basis, we should find some forty thousand scientific societies flourishing in our Galaxy, half of which—older and therefore farther along than we are—will have probably set up communications networks. Our experts should seek out those nearest to us, some five hundred light years away, or five hundred times 6,000 billion miles. In practical terms, this means that a single exchange of communications there and back would take over a thousand years. By comparison, a message to the Apollo astronauts on the Moon and their reply together took only two and one-half seconds.

THE PROJECT CYCLOPS

Radio Waves Vs. Laser Beams

In 1971, NASA and the American Society of Engineering agreed to review the whole question of sounding out intelligent life in the Galaxy, and recruited a staff of researchers from the University of Stanford and the Ames Center in California to do the job. Under the direction of B. M. Oliver, an engineer, the team was charged with working out a new listening system to provide far better chances of capturing interstellar signals than the methods followed since Drake's first experiment at Green Bank eleven years earlier.

The upshot of this initiative was Project Cyclops.

Setting to work, the researchers promptly found themselves in a quandary when this question was raised: aside from the hertzian waves, are there other elements that could carry interstellar signals? Are there different techniques outer space societies could be using for their intercommunications? If so, then radio listening would be a dead loss.

In the vast spectrum of electromagnetic waves, the radio waves occupy only a very restricted sequence. We arbitrarily divide the total range into gamma, X, ultraviolet, visible light and infrared rays and hertzian waves. The spectrum encompasses a whole ensemble of vibratory phenomena which diffuse at the same speed—about 180,000 miles per second in the void—and distinguish themselves only by their frequency, i.e. the rapidity of their vibrations.

Our present techniques enable us to utilize advantageously the least energetic of these waves to transmit information. These techniques are well known: you emit on a narrow band of hertzian waves; on this carrier, carefully maintained and stabilized, by modulation you set the signals the receptors will decode.

In theory, there is nothing to prevent us from using high-frequency waves, such as X-rays, for the same purpose, but the radio astronomers have found that radiations of this kind, coming from very powerful natural galactic sources, are largely absorbed en route to the Earth; by the time they arrive, not much of them remains. Waves of the hertzian type, however, are not subject to such debilitation, at least not during their traversal of space.

For some years now, scientists have known how to render light waves coherent via the laser techniques and use them to transmit signals. This parallel light structure can carry far more information than the hertzian waves. Thus, it is logical to conclude that galactic signals must be conveyed either by laser or radio waves. How, then, do we choose between them?

In 1961, Townes, who introduced the laser beam to the world, first suggested adopting coherent light for interstellar communications. Well before the Apollo astronauts set up reflectors on the Moon to retransmit a coherent light ray beamed from the Earth, from the lunar soil a telecamera aboard a Surveyor had recorded and transmitted weak laser signals of one, two and three watts sent out from Earth; on the photo taken by the Surveyor, the signals appeared as two luminous dots.

This first experiment demonstrated the astonishing directivity of laser light rays. Townes compared the Surveyor photo with others the same camera had taken of New York and Los Angeles by night. Seen from the Moon, the two watts of the laser beam showed up brighter than the hundreds of millions of watts illuminating the two metropolitan areas.

Of course, the distance between the Earth and the Moon—240,000 miles—is piffling compared with interstellar distances, and the Surveyor's photography was still only a laboratory experiment. But since the early 1960s, the laser techniques have made giant strides forward. With gas laser it is now possible to achieve non-stop potencies of ten to one hundred kilowatts (but the power instruments are buried in military secrecy because, unfortunately, they involve research into the "death ray").

Anticipating a surging development of the laser techniques, Townes proposed two systems of interstellar communications based on them. The first would function only beyond planetary atmospheres. In our terrestrial environment, light is subject to considerable distortion through absorption and diffusion. This hampers the laser as a telecommunications medium, and only frequent relays would overcome the difficulty. Otherwise, the light beam would have to be circulated in an enclosed protected milieu, such as a tube.

For receiving galactic signals, Townes suggested an optical telescope with a mirror some sixteen feet in diameter. What he had in mind was the Mount Palomar telescope, then the most powerful light collector in the world (the Soviets have since built a telescope six meters in diameter). This equipment would permit radio astronomers to survey over a distance of ten light years; but fewer than ten stars lie within this range.

Shklovsky didn't think much of this scheme. A signal sent out by such means, he said disapprovingly, would be no more visible than one of the weakest stars spotted by the most powerful telescopes. To make observation possible, the laser emission would have to be far more potent.

Townes' second proposal, involving a ground installation, called for a battery of twenty-five lasers identical with that of his first project, all of them emitting in the same direction and forming a beam of one second of arc. Again Shklovsky came forth with a protest: the stream sent out by such a battery would be inferior to that of the first system. It would appear, therefore, that a laser sender must be installed either in space or on a celestial body without atmosphere like the Moon.

An even more formidable barrier to laser communications, however, looms up from another quarter. Inasmuch as the humanoids we hope to contact will be living in the neighborhood of a star, as we earthlings do, we must find some way to differentiate between the light rays they flash us and the luminosity of their sun. This may seem like a discouraging prospect, but actually, as with radio, the two sources of radiation are marked by wholly contrasting characteristics. A star emits on all the light frequencies of the whole color spectrum and in every direction. The laser beam is directive and shines on a single color. It should be possible, then, to individualize it by coinciding it with one of the sun's strongest spectral rays. These absorptive lines appear in the form of small black bands, in which the light of the laser would stand out as a spectral anomaly. Once this emission is identified, it would be a simple matter to sight intelligent signals through interruption of the abnormal radiation.

But along comes Shklovsky again to demolish our hopes and show us where the snag lies: the laser signal's directivity would prevent the emitter from encompassing the whole target—the sidereal system selected—and the light beam could well miss the inhabited planet (or planets), thereby making the operation useless. A wider beam would require a tremendous boost in power if the signal's indispensable brilliance were to remain intact.

Tallying the pros against these overwhelming cons, the Cyclopeans turned thumbs down on the laser beam and renewed their faith in the Cocconi-Morrison precept: that the physical world offers no means of conducting signals equal or superior to the hertzian waves in the speed of their propagation or the distances they can travel safely.

Tuning in on the Quietest Sky

A matter of primary concern to radio astronomers are the conditions of local propagation. With the atmospheric blanket screening out the greatest part of the radiations coming to Earth, researchers on the ground can intercept a mere fraction of the immense range of the electromagnetic waves; they must be content with detecting visible light and only part of the hertzian waves. Oxygen and the water vapor of the low atmosphere absorb waves of less than one centimeter; the higher electrified strata—the ionosphere—reflect those that exceed fifteen meters. The radio window is but partly open to the sky.

Our gaseous envelope, therefore, is transparent only in a radio band between one centimeter and fifteen meters; it is within this window that communications can be conducted between the Earth and space. But science promises future spatial communications on a far broader spectrum once it is possible to install transmitters and receivers above the atmosphere. Then the restrictions of the terrestrial window will no longer exist.

Meanwhile, taking their cue from Cocconi, Morrison and Drake, the Cyclops researchers have come to realize that the galactic listening window is framed by a certain number of physical imperatives, which must also concern our future correspondents.

First of all, before the radio astronomers can intercept signals, they must be able to distinguish them clearly from the noises of natural origin that jam galactic monitoring. This requires them to operate on frequencies relating to the quietest sky.

Such static varies in type. The Galaxy gives out a background noise which diminishes rapidly as the listening frequency increases—what the physicists call the *synchroton radiation*. These emissions originate in the clouds of dust and extremely rarefied gas that float in interstellar space. The magnetic fields generated by the gigantic galactic rotation trap the cosmic electrons which produce photons.

Electromagnetic signals, on the other hand, have their own background noise, a quantum racket that comes from the granular nature of radiation. It appears on the very small wave lengths, increasing rapidly with the frequency, when the receptor registers the individual shock of the photons of the radiation more and more distinctly. On the millimetrical waves, a signal will be smothered in this quantum cacophony. Thus, the listening horizon is blocked on both sides of the radio spectrum.

Then there is the thermal noise of the apparatus itself and its environment. All matter emits a radiation corresponding to its temperature. This thermal noise is relatively low, about 3° K; nevertheless, since it spreads out over the whole frequency band, it is unavoidable.

These three emissions very precisely delimit the zone of the quiet sky. On frequencies higher than one billion hertz (a GHz), thermal noise falls below galactic noise. Above sixty GHz, the quantum noise will predominate, and the surcharge will mount quickly if we choose still higher frequencies.

But on Earth and other planets that foster life, water vapor and the absorptive lines of oxygen will shut the window under ten GHz. This limitation, however, has no practical consequences in the matter of galactic listening.

In any event, the Cyclopeans recognized the need to choose a much narrower band for their work.

A comprehensive sweep across a spectrum extending over tens of billions of hertz is not conceivable. The operation would require thousands of years. How, then, can the first listening window, too wide for practical purposes, be narrowed? Oliver and his men found the guide line without too much trouble by reverting to the basic equations that give the most favorable countenance to interstellar communications and set forth very general requirements that govern hertzian contacts and the constructional design of antennae, whatever the level of technology might be.

The sensitivity of a radio telescope depends primarily on the surface area of the receptor antenna, a factor which sparked a race for giant antennae over the past few years. A big antenna, however, is badly handicapped in one respect: when it is moved for purposes of orientation, under the heavy weight its receiving lobe becomes deformed. Part of the radiations captured, no longer directed to the focal antenna are consequently lost to observation. Good focalization—a proper concentration of the rays reflected by the big concave mirror—is directly related to the wave length of the incoming radiations. The same need for precision in optical telescopes, as the English astronomer Lord Raleigh calculated, applies no less to radio antennae, which resemble them. Irregularities on the detection surface must be less than 1/16th of the wave lengths received.

The Cyclopeans therefore must envisage giant instruments that will not tolerate a choice among wave lengths that are too short. The deformations on an antenna meant to function on twenty-one centimeters, for example, may not exceed two centimeters. No technological improvement in the antenna will do the trick; the malleability of matter has its limits.

To assure the best performance of the biggest possible antenna, the working frequencies must therefore be relatively low. Oliver's crew must turn to the longest waves of its first window on the quiet sky, between one and two GHz.

Since the choice was dictated for fundamental reasons, Oliver declared, the logic of this tuning-in limit should be valid even at a thousand light years from the Earth. It is here, between one and two GHz, that the Cyclops researchers will find the best conditions for reception, the least cosmic static and the most economical monitoring.

In this part of the spectrum reduced to one GHz, where

interstellar communications are most propitious, what still remains to be done is to settle on a much more modest frequency band of exploration; obviously it is not possible to locate an artificial signal, necessarily very narrow, on a one-billion-hertz band. This would be no simpler than hunting for that needle in a haystack.

Narrowing this overextended window is no longer a question of invoking some physical imperative but rather of deciding on a listening band without guidance either from nature or technics. But this freedom of choice, which our galactic friends must share with us, is disquieting in one very real sense: we may never find the contact.

The Water Hole

From 1959 onward, as we have seen, Cocconi and Morrison urged the choice of the hydrogen wave length, set in the window of the short waves and, because of its predominance in the cosmos, presumably a universal standard. This privileged emission, they argued, could not escape the attention of an advanced society, which would certainly assign it a special importance. We can therefore surmise that outer space technicians, limited in their choice of an emission frequency, must have undoubtedly tried the hydrogen frequency.

The hydrogen wave length, however, is beset by one serious disadvantage: it is accompanied by so much static that a twenty-one-centimeter emission can be drowned out by the glut of natural background noises. Consequently, some radio astronomers have suggested shifting the wave length somewhat from the natural standard to a multiple or a simple fraction. A parallel frequency—for example, half or double the hydrogen wave length—would make listening easier while still preserving its kinship with the number one element of the cosmos. Nonetheless, fervent partisans still champion the hydrogen standard. Project Cyclops has, of course, mulled the question and come up with some interesting conclusions.

For lack of physical imperatives, Oliver observes, nature seems to wink an eye at us by way of discreetly indicating a listening range for a privileged frequency.

The frequency characteristic of hydrogen lies at 1,425 MHz, which is near the middle of the window, with a scope ranging between one and two thousand MHz.

At 245 MHz higher up the scale of frequencies, we find the first hydroxyl ray, a fusion of two atoms, oxygen and hydrogen. (Radio astronomers find water, H_2O, still higher in the frequency spectrum.) The quietest sector of the whole spectrum lies between these hydrogen and hydroxyl rays; no other emission has ever been detected between these limits. The Project Cyclops team thinks that this band, situated between the two radio-electric resonances of two water constituents, is the one definitively marked for interstellar communications. Indeed, what zone could be richer in significance than this "water hole" so closely related to the indispensable support of life, to the universal solvent essential to the evolution of all organisms, to the environment which has generated all living beings?

The Cyclops report proposes to guarantee this silent band against interfering emissions, just as the hydrogen wave length has been patronaged for the benefit of the radio astronomers. This would mean, in short, extending the protection of some 200 MHz as far as the first line of hydroxyl to safeguard this galactic window against any jamming by terrestrial emissions. Meanwhile, certain frequencies within this listening range have already been utilized, mainly for the needs of navigational satellites.

In a word, nature has favored us by reserving a first-rate listening window on the Galaxy. In the aperture of this window, which opens out on the quietest sky, she graciously suggests that we select the band separating the hydrogen and hydroxyl wave lengths because it offers the best listening technique and, in all probability, the best prospects for tuning in on other races in other worlds.

How Will They Make Contact with Us?

Having chosen what they consider the frequency band most likely adopted by their outer space correspondents, the Project Cyclops experts must then turn their attention to the study of a suitable antenna.

The most sensitive receivers pile on considerable background noises to the static caught by the antenna. The apparatus multiplies by five the "noise temperature" (of 4° K) it receives from space. An ideal receiver, then, must multiply by five the potency of the signal expected so that it will emerge clearly above the static din.

Moreover, the power of emitted waves diminishes with distance, according to the law of the square. The collective antenna will achieve the required sensitivity if its surface is expanded; the bigger the antenna, the more energy it will acquire. What, then, should be the dimensions of the Cyclops antenna?

Oliver admits that he can give only an approximate reply to this question because he does not know the extent of the radius in which to search for an eventual signal; nor does he know its power. Conjectures on the number of our galactic partners who can engage in telecommunications are extremely uncertain; and it is difficult to surmise the duration of their contact efforts.

Von Hoerner separates galactic signals into three distinct categories: the radio-electric emissions caused by industrial operations and radio-television transmissions—purely local pursuits which have nothing to do with outer space contacts; contacts purposely directed, presuming the existence of telecommunications established with other worlds or, at least, a contact established with one other world; and call signals sent out to attract the attention of other outer-spacemen prior to an actual transmission of information. The Cyclopeans have taken due note of these three classifications.

Shklovsky maintains that a local radio-electric emission must be totally indiscernible to any listener-in ten light years away (in the zone of our nearest stars). Such an industrial radiation would have to be hundreds of times more potent—a possibility not to be excluded in the future when man will have learned how to exploit the essential sources of energy in our solar system.

Nor do the signals of the second category—long-distance calls—have much chance of being heard by our earthly antennae. Von Hoerner has reckoned that such interceptions would be possible only if each advanced galactic civilization were to converse simultaneously with some 1,300 neighbors, which is hardly likely. Our chances of hooking up with an interstellar communication network already in operation are by no means promising.

The most hopeful prospect, then, will be the third category, non-informational call signals. They must be economical and able to penetrate enormous distances at a minimum expenditure of energy; hence, they must adopt the narrowest possible emission band and a precise frequency which any prospective listener will already know about. A unidirectional band consumes far less energy than one beamed in all directions. It is this factor that will enable us to cope efficiently with the solar emission when we begin transmitting toward the Galaxy. It is also the reason why emitters installed on distant space vessels—planetary sondes—or on the Moon are equipped with high-gain antennae pointed to the Earth so that they can transmit at the nominal cost of a few watts.

The Project Cyclops researchers are banking on this possibility.

To the pessimists, supposedly no permanent contacts yet exist among the galactic societies within our reach. These societies have not gone beyond the stage of calling and listening. If this is true, then we must prepare to launch a comprehensive search for signals over a range of one thousand years.

In the optimistic view, however, outer space civilizations are already in contact and presumably attempting to extend their network by enlarging its sidereal membership.

Oliver and his aides reason that the "communications societies," already in the process of organizing a chain, are casting about to enrol other worlds. We should then have some opportunity of observing their signals within a radius of less than one hundred light years. Within this spate, we should also be able to capture traces of their activities not particularly meant for our ears—the residual radiations, of which Kardashev and Dyson spoke.

On these considerations, the Cyclops group has formulated a basic model of probable galactic connections. Our researchers think that their eventual counterparts in outer space have very likely been led to equilibrate their transmitting and reception methods on the intersidereal scale. Such an equilibration is in the nature of things; like policemen patrolling their beat in pairs, transmissions go by twos. Furthermore, the theory of antennae prompts us to choose this pattern since it identifies those antennae which emit and those which receive. Considering the possibilities of distances, the Cyclops team calculates that an emission power of one thousand megawatts (one billion watts) is not unreasonable.

This figure may seem wildly extravagant, but one must bear in mind that here Oliver is falling back on the hypothesis of an omnidirectional source of radiation (an astronomical reflex; the stars radiate in all directions). In fact, a directive emission via a parabolic antenna would utilize only an infinitesimal fraction of the power beamed in all directions to channel the same amount of energy toward the Earth.

In terms of this probabilistic theory, Oliver and his men figure that to intercept a signal of this type originating in a point one thousand light years away, they would need a giant antenna measuring several miles in diameter. Such a construction is, of course, out of the question.

On the ground, observation instruments—which must be mobile—cannot exceed a certain mass; we have seen that they become deformed and unmanageable. Therefore, some compromise solution must be found, such as a fixed structure with a mobile collector focus, such as those at Arecibo and Nançay, in France. But the pointing of these apparatuses and their potentialities are too limited to meet the demands of Project Cyclops.

On the other hand, an immense radio telescope assembled in space is not inconceivable. Empty space and weightlessness permit satellized engines to rotate very easily. But they must be properly stabilized. These techniques have achieved a remarkable degree of refinement, especially in communications satellites in orbit some 22,000 miles above the ground, with their directional antennae permanently aimed at the same sector on Earth and their solar panels to the Sun.

The Cyclops team recoils in dismay at the exorbitant cost of such a spatial operation. Since retrievable transport spacecraft do not yet exist, the construction, installation and operation of such titanic equipment would require a prohibitive outlay of money. Heavy observatories in orbit above the atmosphere's befouled windowpane will only be possible when NASA can build recuperable space shuttles—toward the end of the 1970s.

Listening to a Million Stars

Obliged to forego an antenna several miles in diameter, the Cyclops team came up with another scheme, more practical and endowed with other advantages. Instead of a single, gigantic collecting surface, this system foresees a whole array of radio antennae, numbering from one to ten thousand, each sixty feet in diameter. Building and operating units of this dimension pose no major problems that the experts do not have under control.

In point of fact, it turns out that such a pluri-antennae system would be even more effective than a single colossus because it allows for more precise pointing. Since each antenna could be oriented at will, the directivity of this method would prove to be much finer.

In studying this project, Oliver and his aides concentrated on the instrument exploiting the collected waves, which involved meticulous problems of regrouping the flow intercepted by the different antennae and carefully preserving the figure of each original radiation, inasmuch as the same phase relations must be scrupulously maintained; the signals carried on a working frequency must necessarily be transmitted to the central collector in the identical form in which they are received. A harmonious operation of the various receivers, control over them and the work of recording and measuring their signals would be automatic, with the entire sequence of events under the permanent command of computers.

A network of antennae hooking into an efficient receiver could be built over a period of ten to twenty years. All taken together, every thousand or so of them would cover a total surface of twelve square miles. The research work could begin during the course of construction with eavesdropping on the nearest stellar systems, and develop as the network spreads. The directive benefits provided by each element would eliminate any blind search of the whole sky or group of stars. The radio astronomers would be obliged to eavesdrop on the stellar systems one by one.

The candidate stars, those escorted by planets on which life is possible, lie in the "principal series." Presumably they belong to solar type G and the stars of the near categories (end of type F and the beginning of type K). Within the span of one thousand light years, their number is estimated at one million. We do not yet possess a catalogue of them, but before any such research program can take off, these one million stars must be identified and logged by optical observation. The Cyclops report suggests how this preliminary work should be organized.

Oliver counts first of all on a prospecting program conducted within a smaller sector, several hundred light years away from us. This minimal program would conform to an "optimistic hypothesis." In this relatively closer sector, we can count about one thousand candidate stars. If the Cyclops researchers were to explore only this zone, they could dedicate several hours or even whole days to tapping each one of them. Thus, they could look out not only for voluntary signals but also for radio-electric emissions giving information on the local planetary activities of intelligent beings.

But the experiment presents one extremely difficult problem. Since the Cyclopeans decree that the projected monitoring time must not exceed thirty years, no more than one thousand seconds could be allotted to each of the one million stars thought to be capable of fostering life in its environs.

A listening spate of only one thousand seconds is indeed very brief. The monitor would necessarily have to limit himself to the business of detecting steady signals and exclude isolated calls, which could easily pass unnoticed.

For optimum receptivity, the radio astronomers would be confined to a very narrow frequency band. To cover the selected spectrum of two hundred megawatts would require multiplied listening on a channel of 0.1 hertz.

Even working in such economic conditions, convenient for long-distance detection, the Cyclops researchers would be glued to their sets for three hundred years in order to auscultate a single star in a 100-megawatt band, corresponding to only half the width of the "water hole" retained as the field of exploration.

The grizzly problem, then, is to reduce the search time from three centuries to one thousand seconds, if the exigencies of the program are to be fulfilled.

The American technicians have actually hit on a method to realize this drastic time reduction—by exploiting the resources of opto-electronics.

This is not merely a matter of sweeping over a wide band at high speed to find a coherent signal; the Cyclops system must also command the facilities for analyzing in depth the noise intercepted. The radio astronomers have noted that apparently shapeless electrical emissions recorded on their sets are actually formed by the superposition of simpler signals which correspond to distinct emissions. Jean-Baptiste Fourier has given us the mathematical formula for these transformations.

Thus, if the complex noise intercepted by the antenna could be analyzed and separated into its varying elements, the radio astronomer could single out the modulation of a hidden signal.

For an instantaneous analysis of the whole radiation flow intercepted, the Cyclops team would modulate a coherent light beam by means of signals caught by their antenna. Put briefly, the modulated laser ray supplies a model of the captured complex signal. It would be a simple matter, as the listening program advances, to film the variations of intensity recorded on a wide part of the band monitored, then to follow up with an instantaneous optical analysis of these sequences through another laser beam and a lens which, invested with convergent power, could portray the complete spectrum of the emission received. A single analyzer would cover one megahertz band width in one thousand seconds. Two hundred analyzers could embrace the whole spectrum of the "water hole", leaving only a regrouping of the entire band to be done. A sweep of the raster would make it possible to ferret out the series of registered pulses which indicate intelligent signals; all the impulses of the periodic signal would be added in phase and, in the process, the signal would be sifted out from the surrounding noises.

Such a sweep, however, must be carried out in various positions because the signal will show up on a drifting frequency; it will come on a slant. This Doppler drift can result from several causes: the radial speed of the star aimed at, the movements of the planet sending out the emission or the movements of our own globe, which can affect no less the frequency of an outer space signal.

Comparing the vast Cyclops mechanism with the equipment Drake used in 1960, we find that the former would be

200,000 times faster in its analysis of the radio-electric spectrum, and its sensitivity ten million times superior. Drake's eighty-five-foot antenna pales into insignificance before the total, massive Cyclops battery, equivalent to a collector instrument with a diameter of 16,000 feet. The Cyclopeans have succeeded in reducing the temperature noise (the thermal interference emission recorded by the receiver) from 350° K to 20° K. The reading time required for a unit of information drops from one hundred seconds to only ten. Most important, the program in its entirety is infinitely more ambitious, with the explored spectrum expanding from one hundred hertz to two hundred million.

No less spectacular is the difference between the financial requirements of the two experiments. Whereas Project Ozma got off lightly with a mere million dollars, one blanches at the prodigious cost of realizing Project Cyclops: ten billions of dollars, or approximately half the budget of the Apollo program.

Oliver fears that the most formidable obstacle that could block Project Cyclops is the uncertainty of the ultimate results it would produce. All the same, the program merits serious consideration for several reasons. Its structures could be built on a long-term basis and put into operation in progressive stages. A modest, initial network of the antennae designed for it would proffer a very efficient interferometer. Its extreme sensitivity and directivity would invest it with a priceless value as an observation instrument for the wave lengths it is intended to monitor. In short, it would pave the way for tremendous strides forward in astronomical research.

Moreover, the system would help immeasurably in surveying space. We must await the development of astronautics capable of very long-range action, demanding a far more ambitious network of telecommunications than we now possess. Pioneers X and XI will be beyond our reach before they cross into Pluto's orbit. In an early stage, it would maintain contacts with spacecraft flying in such remote domains. These sondes, in fact, transmit on a decimetrical wave length like that which the Cyclopeans believe would bring forth optimum results from their antennae.

What the Soviets Are Doing

With its proposed mastodon galactic-scanning installations and instant data-processing methods, Project Cyclops is scheduled to operate within the time limit of one generation. A longer experiment would be no more practical than one begun with hot-air balloons in the 1820s and still going on today.

Meanwhile, what are the Soviets up to?

Kardashev has various new listening-in strategies in mind, mainly one based on the simplest kind of nondirectional reception—a dipole antenna exploiting the frequencies least disturbed by cosmic noise, those on the decimetrical waves; the test, however, could be extended over the whole spectrum of radio waves. But picking up static from all directions, this system would be nondiscriminatory. Kardashev, therefore, suggested two different receptor sites to function simultaneously, a dual device to determine whether the intercepted signals originate in outer space or within the Earth's atmosphere. In his view, this technique, albeit unpretentious, should make it possible to individualize a signal on an electromagnetic impulse. Some natural emissions, like those from supernovae, could serve as a simple signal model. These stars in rapid expansion are subject to series of shuddering explosions, each one reacting on the magnetic field and generating a pulse on the radio wave spectrum, detectable on Earth.

Kardashev has also demonstrated how a sizable directional antenna can secure optimum results in a monitoring project oriented to the center of the Galaxy. If the beam of such a radio telescope is adjusted to the angular diameter of the galactid hub (a two-minute arc), the surrounding cosmic interference that disturbs less directional observations would be eliminated at a single stroke. Under these conditions, the best listening should be on the five-centimeter wave length, for which Kardashev recommends an antenna 625 feet in diameter.

Addressing the Byurakam conference, Professor V. C. Troitski gave details of other experiments which had been attempted in the Soviet Union. One of the first was a program aimed at finding monochromatic signals—radio signals on a very narrow band width, like our terrestrial emissions. This began with the scientists drawing up a catalogue of eleven nearby stars within a range of a few light years, each of a spectral type like the Sun, signifying that they are approximately coeval and could have had a development parallel with that of our solar system. The researchers assumed that the call signals they were looking for should be of the purest sinusoidal kind, which stamps artificial signals. Anticipating that results would be slow in coming, they opted for a very narrow reception band and chose the twenty-one- and thirty-centimeter wave lengths for the standard reasons we have previously stated. Using a receiver capable of scanning a two-MGh band in ten minutes and a thirty-three-foot antenna, in Gorky in the autumn of 1968, the team listened several times to each of the eleven target stars, but caught no monochromatic emissions.

In addition, a six-session monitoring experiment tuned in on the Andromeda Nebula, 1.8 million light years away. Kardashev thought that monitoring an extra-galactic system would appreciably increase the likelihood of establishing a contact because it means surveying innumerable sidereal worlds simultaneously, all of them capable of supporting intelligent life. Calculating the cost in energy of an emission from Andromeda to our Galaxy, he concluded that a type III civilization could afford the luxury of a radio beacon transmitting over a distance of two million light years although, since it would take two million years for a signal to travel from one galaxy to another, such a feat could only be achieved by extremely long-lasting societies. Shklovsky wondered whether the transmitting society would still be around by the time its signal reached its destination; many intergalactic communications could now be only archaeological remnants of defunct populations. In any case, the parabolic antenna at Gorky could not record signals from any civilization preceding our own by eighteen thousand centuries.

Current Soviet research, begun late in 1974 at Zelentchouk in Caucasia, uses a giant circular telescope 600 meters in diameter, exploitable on the millimetrical waves (best used on the eight-millimeter wave length), which began operations at the end of 1974. It was built according to data supplied by an experimental model installed at Pulkovo with one section of arc. This radio antenna should bring over 1,000 stellar targets within the Soviets' accessible range.

Other experiments are going on in the U.S.S.R. In one of his projects, Kardashev put radio astronomers to work to locate, if possible, the background noises issuing from some technological society's industrial activities. Marshalling the full decimetrical band, they tested simultaneously in several sites on Soviet territory, figuring that parallel monitoring should make it possible to distinguish between remote signals and local emissions. The first trial began in March, 1970, in Gorky and the Crimea, some 1,500 kilometers apart; the second, in November and December of that year, employed four observation posts, in Gorky, the Crimea, Murmansk and a site on the Ussuri River in Siberia for a deployment of over 8,000 and 15,000 kilometers. Each station was equipped with a dipole antenna and a reflector pointed to the zenith. The main problem was to sort out local noises from galactic signals by correlating the emissions heard at the four stations.

The results of these tests were negative: the only signals recorded were related to solar activity and came from the atmosphere. The Soviets consequently agreed that it would be wise to look for sporadic emissions on short waves between ten and fifteen centimeters, where confusion with natural emissions is not very probable.

Like the Americans, the Russians took steps to integrate the search for intelligent signals into their overall radio-astronomy program. In Caucasia and the Russian sector of the lofty Pamir plateau in West-Central Asia, the Soviets set up a detection system connecting two listening posts 3,000 kilometers apart, both equipped with highly sensitive omnidirectional receptors and situated in relatively quiet areas not too much disturbed by normal radio traffic.

In October, 1973, Professor Samuel Kaplan, of Gorky, announced that research into radio waves of galactic origin was proceeding with more than thirty thousand Soviet specialists working under the direction of Professor Troitski. The project relied on four widely separated points to isolate cosmic signals from local interference. Kaplan notified the Tass news agency that the network had picked up radio emissions from beyond the Earth, which could have originated in the upper reaches of the atmosphere, but Troitski thought they could also have come from some extra-terrestrial point.

To his sorrow, however, it was soon discovered that these intelligent signals were being emitted on the centimeter band by an American satellite of the Big Bird type, which had been launched from the military base at Vandenburg, California, 13 July, 1973, carrying aerial radar reconnaissance installations, which can penetrate the clouds, whereas cameras cannot. First launched in 1972, the Big Bird satellites were the first to use a radar detection system directed to the Earth's surface. (Techniques of the same kind were adopted on the last Apollo flights to scan the lunar surface.)

Two other Soviet teams have been furthering the detection program. One, using the Murmansk, Gorky and Ussuri sites plus an observation platform aboard the oceanographic vessel Akademik Gourtchatov, the widest possible base, aims to locate with maximum accuracy the different sources recorded by the radiometer. The second group utilizes movable equipment in Caucasia and on the Pamir plateau—remote sites offering the advantage of high-altitude observation and freedom from industrial interference, a growing form of space pollution which is making reception of cosmic emissions constantly more difficult.

Now the Russians are planning to widen their monitoring base from Caucasia to Kamchatka, a stretch of 8,000 kilometers, and between the Siberian Far East and the French telescope at Nançay, giving them an observation base of 10,000 kilometers.

From Arecibo to Messier 13

In the United States, other experiments soon followed Drake's historic effort. Two attempts at detection were made with the help of the 300′ and 400′ antennae at Green Bank, in the course of which the scientists remained faithful to the hydrogen channel which is the wave length most often used by the institute. Neither project was confined to sinusoidal signals.

The first experiment concentrated on ten nearby stars selected from within a radius of seventeen light years, of which the nearest is Barnard's Star. Any one of these could be capable of supporting life.

The second and more ambitious of the two attempts covers a catalogue of two hundred stars. If results up to now have been negative, this is hardly surprising since it is estimated that, in order to have a good chance of detecting an extra-terrestrial civilization, we must turn our receivers to one million stars. Moreover, this is an optimistic estimate which presupposes a galactic population closer to Sagan's million than von Hoerner's postulated forty thousand.

We know that statistical projections of the galactic

population do not encourage short-range listening. If we do have near neighbors, they must be civilizations with a lengthy potential life-span, since longevity is the factor on which population density in our Galaxy principally depends. According to Kardashev's hypothesis, the civilizations which last longest are those which are distinguished by expansion and by their mastery of the techniques of galactic communication. In other words, intercepting their emissions should have presented no problems for us.

This means that we must revert to the second of our two feasible hypotheses which postulates more scattered populations, societies which die out more quickly, a more modest level of technological advancement, and hence implies that we must encounter greater difficulties and penetrate much further into the galaxy before we find beings with whom we can communicate.

It is of course true that the statistical distribution theory is just that—a theory. It may well be that radio astronomers will discover beings similar to ourselves at a more modest distance. But we must in any case refrain from attempts to estimate the success potential of such research. It is precisely this kind of detective game which Project Cyclops was designed to avoid.

Meanwhile attempts continue to be made in the United States. In 1971 and 1972, receivers were tuned periodically to the two nearby stars which figured in Drake's 1960 experiment, and, in the opinion of the researchers, Tau Ceti and Epsilon Eridani still rate high on the list of candidates. Van de Kamp has made the point that the disturbances to which their course is subject may mean that they are accompanied by satellite planets, but he is not in a position to give more exact details.

Recently the scientists have begun to utilize detection techniques other than radio-wave monitoring. In 1975, NASA launched its space observatory Copernic to search for intelligent signals in the form of ultraviolet emissions on the supposition that extra-terrestrials might be transmitting on ultraviolet ray beams. Herbert F. Whischia, one of those in charge of the program, declares that while the natural emissions of stars like our Sun are relatively slight on the ultraviolet wave lengths, a coherent artificial ultraviolet signal of the laser type originating in the vicinity such a star would run no risk of fading out in the overall light radiation, as would happen if it were emitted as visible light. Transmission by ultraviolet laser beam, therefore, offers a feasible channel of interstellar communications. But radio waves claim a considerable advantage in very long-distance communications in that they are less likely to be absorbed by interstellar matter than beams radiated on a higher frequency, although ultraviolet techniques are still practical for contacts among neighboring stars. On this premise, the Copernic scanned ultraviolet emissions in the vicinity of Epsilon Eridani, Tau Ceti and Epsilon Indi, about eleven light years away.

No longer content with waiting to hear from outer space, American radio astronomers are now sending out signals of their own.

Over its antenna, 1,000 feet in diameter, the Arecibo installation has been transmitting periodic call signals to Tau Ceti, about 12,000 light years away. This giant radar, built by Cornell University, was originally designed to penetrate a distance of 6,000 light years on a 100-hertz passing band, but radio astronomy quadrupled its performance potential within the span of one decade, leaving it behind the times. The dish was recently resurfaced with a perforated aluminum cover; and a more powerful transmitter was suspended above its center. Radar sensitivity has been increased 2,000-fold, and the frequencies that can be observed for radio astronomy have been expanded tenfold.

This done, in November, 1974, Drake launched an ambitious experiment, choosing as his target a cluster of about 300,000 stars on the rim of our Galaxy, 24,000 light years away, known as Messier 13.

Heeding Kardashev's recommendations for far-distant communications, Drake adjusted the huge antenna's

ARECIBO MESSAGE

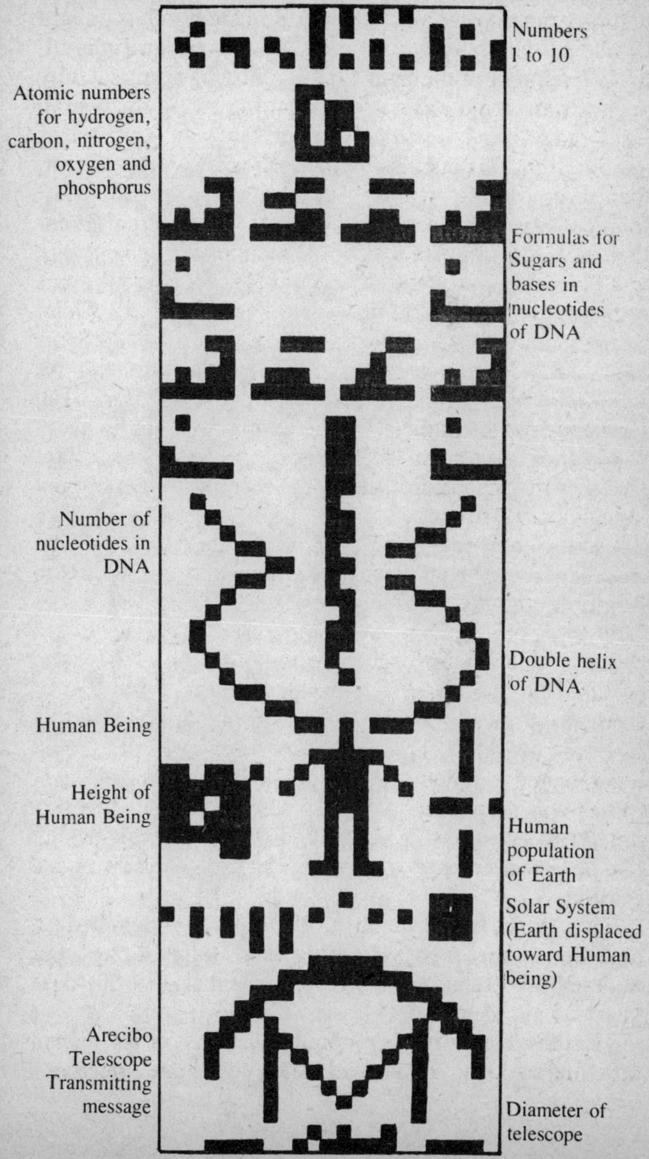

directional beam in such a way that it would blanket the entire target and reach all the populations presumably living there. To secure the antenna's optimum performance, he chose the 12.6-centimeter wave length.

This time, Drake was not satisfied merely to transmit a call signal in the hope of striking up a contact. This would have been futile since the distance is so great that a reply would only be forthcoming in 48,000 years. Instead he preferred to transmit to Messier 13 an extensive catalogue of information on the Earth's biochemistry. To make sure that his message had a reasonable chance of reaching one destination or another in that far-off realm, he sent it out by shifting the binary number code between two radio frequencies, so that both types of signals could be identified accurately, should they be intercepted. Elaborating on the hypothetical message he had composed to test his colleagues at Green Bank years before, he tapped out a series of 1,679 signals within less than three minutes. Any outer space radio astronomer, according to Drake, would understand that the binary digits can be broken down into 73 rows of 23 bits each, 1,679 being the multiple of these two primary numbers.

Printing the "ones" as black squares and the "zeros" blank, a picture emerges. This visual message can easily be interpreted.

The first line gives the code from which the recipient can work out a numbered sequence (from one to ten), and he will see that these sequences are meant to be read from right to left.

Deciphering the first rebus, he will make out the atomic numbers from the five basic elements in our body chemistry: hydrogen, carbon, nitrogen, oxygen and phosphorus (from right to left: 1, 6, 7, 8, 15). Continuing, he will learn the molecular structure of various sugar compounds formed from these five elements, and discover four diagrams of the phosphate group (one atom of phosphate and four of oxygen). Drake's intention is to enlighten his Messier 13 auditor on the structure of DNA, our body chemistry's basic acid, vital to the transmission of the genetic code, which perpetuates the human race. He will then discover a graphic representation of DNA with its double helix structure and sugar-phosphate chain.

Now the report sketches an earthling, giving his measurements in multiples of the transmission wave length—14 x 12.6 = 1.764 meters. To the left of the figure, another code sequence indicates our global population, about four billion. Below this, a diagram sketches the solar system, with its nine planets, the third of which can be identified as the habitat of the race sending the message. There is also a design of the Arecibo antenna, with its dimensions given below, and a clear indication that it transmits by reflection from a parabolic surface.

This three-minute message took five hours and twenty minutes to pass beyond the outermost planet, Pluto, and leave the solar system, a brief moment in the 24,000 years it will be travelling on its way to Messier 13. If we accept Sagan's estimate of one million progressive societies in our Galaxy, one or two among the 300,000 stars in that cluster might sponsor civilizations which are up to receiving and interpreting Drake's message. But will their scientists know how to pinpoint our Sun from the myriad of stars in the firmament? Perhaps. Seen from far out space, the radio signals that emanated from Arecibo are as bright as ten million suns. They should not pass unnoticed.

LOOKING AHEAD

Jail Break

While conceding the probable existence of other intelligent worlds, scientists have still been sorely censured for their conservatism, their lack of imagination in casting these societies in much the same mold as the Earth model and their lack of vision in contemplating viable means of entering into contact with them. Those who scold them insist on a prickly problem already covered in this book: is it not likely that our silent friends in outer space might be following communication techniques other than those our radio astronomers cling to so persistently?

Although along with Morrison and Drake, Cocconi gave top marks to the hertzian techniques, yet at the outset of his investigation he himself pondered the feasibility of other types of communications carriers. In particular, he turned his sights, if only momentarily, on the gamma rays, located opposite the radio waves in the electromagnetic spectrum.

Cocconi, a specialist in this kind of emission, worked with the European Organization for Nuclear Research in Geneva, Switzerland. He had studied the Crab Nebula,[I] a

[I]The Crab Nebula consists of the remnants of an exploded star, first observed by a Chinese astronomer in the year 1054. These vestiges, formed mainly by a cloud of particles in expansion, emit radiations on very different wave lengths, from gamma rays to the rhythmed radio waves, like pulsars. Study of the Crab Nebula is one of modern astronomy's principal pursuits. It has yielded invaluable information on the evolution of the stars and the Universe.

source of gamma rays well known to the experts, on the theory that galactic populations might have sought to prevent their signals from being drowned out by adopting some message-bearing wave that could be differentiated from the radiations of their host suns. Two types of emissions meet this condition at the opposite extremities of the spectrum: the radio waves and the most energetic gamma rays, those on the highest known frequencies.

Cocconi thought that the hard gamma rays, so sparse in the sky that they are clearly perceptible, could serve as ideal message bearers, but they struck him as much less ideal when he began to tot up the exorbitant cost of such a system. To produce these waves requires a synchroton, one of those massive constructions worth millions of dollars, used in nuclear research centers. Transmission by hertzian waves is so much more economical that he saw no choice between them. To this stripe of Western reasoning Shklovsky retorted, not without good reason, that the questions of economics in an earthly contingency was futile when other worlds might have resolved them in altogether different ways, if they have to cope with them at all.

Whatever the electromagnetic wave we might use, the central problem of a suitable potency still remains to be tackled. If the experts take a hard look, they might well scratch their heads and wonder whether they haven't imprisoned themselves within the confines of their special techniques and knowledge; if so, they must plot a jailbreak. Von Hoerner draws an analogy between this constrained situation and the ideal zoo, in which the animals are free to wander about in a happy illusion of absolute liberty, spared cages and caretakers. Pursuing the communications techniques we have thus far developed could indeed entrap us in a similar, rosy illusion that, like the animals in von Hoerner's easy-going zoo, we are operating in an aura of untrammelled independence. In reality, we rely upon the electromagnetic waves entirely for our understanding of the galactic world. They are the media that permit us to measure the infinite stretches of the Universe, but they also impose severe restrictions on our information, they are the invisible bars of our prison. Perhaps, however, we can remove them and escape to freedom.

Let us examine the question more closely. The physicists associate the electromagnetic waves with photons, those grains of energy which, having no mass, can circulate at tremendous speeds, somehow materializing the information brought to us by the waves. (We must accept the incomprehensible fact that waves have no matter themselves, only energy, but join with the microscopic unit known as the photon.) Aside from the photon, are there no other particles that can also carry signals? The Project Cyclops team, all experts in radio-astronomy and in telecommunications, turned down this possibility. Yet Richard Arnold, a physicist at the Argonne National Laboratory of the University of Chicago, performed a singular contact experiment with muons. [This particle, which disconcerts the physicists, is a sort of elephantine electron that appears in reactors when atomic cells are smashed by bombardment.]

The same reaction occurs on the threshold, or upper rim, of our atmosphere under the impact of primary cosmic rays. We note then that when the bombarded cells disintegrate, they release energy in the form of particles known as mesons pi, the materialized substance that glues the cell's neutrons and protons together. But transformed into grains, this cohesive agent instantly loses its stability. In less than one-millionth of a second, it will pulverize into two other mysterious particles, the muon and the neutrino. Reactors in subnuclear research laboratories produce them in vast quantities. Since they easily penetrate the atmosphere without being screened out by the Sun, Arnold has suggested that they could transmit signals.

An initial experiment was undertaken at the Argonne laboratory in April, 1972, when a message in Morse code was transmitted through a cement wall to a receiver five hundred feet away, duly intercepted and decoded. The receiving apparatus included two scintillation counters. Arnold declares that muon communications would compete with the techniques now used. But muon beams disintegrate too rapidly, and their effective trajectory goes no farther than five hundred miles. By all the evidence, then, these particles must be excluded.

The neutrinos, on the other hand, are extremely stable phantom particles, practically without mass, which circulate at the same speed as the photons. They have almost no reaction to matter; moreover, they are universal. Most of the neutrinos that penetrate the Earth are produced by the heat of the Sun and the stars. They traverse the Galaxy; we are bathed in their fabulous flow. Unfortunately, the rarity of their reaction to matter makes it extremely difficult to spot them. They are electrically neutral and they do not react to strong force. While our earthling scientists have not yet devised an efficient neutrino detector, undoubtedly some outer space civilization has already come into possession of such an instrument. If so, then it is possible that its technicians are trying to signal us by releasing muon particles on a time schedule, much like American Indians relaying messages by smoke signals.

Although we are far behind in this respect, all evidence points to the very real possibility that the neutrino could dethrone the photon. Arnold himself believes that his experiment with a flow of muons will pave the way to future exploitation of neutrino beams.

If some means could be found to ferret out easily this elusive particle, man's research would progress by leaps and bounds. Once put under observation, the neutrino would enable our scientists to discover the world of anti-matter in the Universe, and perhaps anti-galaxies. The electromagnetic signals, the only ones coming to us thus far from the great cosmos, are useless in this area because the photons in their radiations are their own anti-particle. An anti-galaxy must radiate the same type of photons as a galaxy does, therefore we have no way of knowing whether our Universe consists of worlds of matter or worlds of anti-matter, as some persons hypothesize.

It is even possible that this huge Universe is formed only of a symmetrical tissue of neutrinos and anti-neutrinos, apart from insignificant local occurrences we know as matter.

Even if it fills a lesser place in the fabric of the Universe, the neutrino offers the observer an excellent opportunity to break the photon's monopoly over our information. If galactic civilizations superior to ours do in fact exist, they will most likely have acquired a knowledge of the neutrino and perfected new techniques based on it. Practising neutrino astronomy, they will have profited by its support and utilized it for establishing contacts with other worlds.

Apart from electromagnetic waves and mass-less particles like the neutrino, some persons are convinced that still another method of communicating with intelligent extraterrestrials could be mental telepathy. Among them is the astronaut Commander Edgar Mitchell, LM pilot of the Apollo 14 moonflight who, with Alan Shepard, landed on the lunar surface in February, 1971. It is his belief that estrasensory perception might prove to be a more effective means of linking galactic societies than radio astronomy.

An enthusiastic practitioner of telepathy, Mitchell took advantage of the occasion to conduct a series of long-distance ESP experiments during his flight to and from the Moon with the collaboration of partners on the ground. Programmed to be held during his rest periods, these tests were not included on the official agenda of the mission.

During his long training period prior to the flight, Mitchell found no opportunity to prepare and rehearse them systematically. Before the blast-off, however, he did manage a trial run, for purposes of comparison. Choosing four colleagues as recipients of his mental projections he concentrated on a series of numbers printed on the reverse side of his check-list as a substitute for the Zener cards generally used in experiments of this kind (astronauts are not permitted to carry any non-essential material on board). But during the flight, various unforeseen incidents upset the timetable prearranged for the experiments, so that analysis and interpretation were undertaken in circumstances far more difficult than the participants had expected.

Three separate teams, following different analytical methods, examined the results and agreed that they were positive, although failing to achieve a level high enough to be considered truly significant. Since it appeared that Mitchell's subjects received his messages before he transmitted them from Apollo 14, the tests were thought to be more a matter of precognition than telepathy.

The phenomena of telepathy and precognition are only now beginning to find scientific confirmation. At Maimonides Hospital in Brooklyn, for example, experiments conducted on subjects under hypnosis have testified that such inexplicable mental communications do in fact exist.

Mitchell's tests lent no substantial support to the hypothesis that information can be imparted by ESP, yet it may be possible eventually to perfect this technique into a valid communications carrier. Soviet teams, too, have experimented with long-distance ESP communications between European Russia and Siberia, but there is no evidence that they have duplicated them in space. In any event, for the present at least, ESP is no match for radio contacts.

Can We Understand Ourselves?

The hypothetical example of the neutrino shows that galactic societies can communicate by various means, depending upon their stage of development; but our ignorance of their evolution makes it difficult to select a common contact.

Pierre Guérin, a French astronomer, states the problem very clearly: there is no way of forecasting with any accuracy what our science will be like several centuries hence. We persistently underestimate the potentialities of innovations awaiting us in the future. A gap of thousands of millions of years between galactic societies must certainly result in a plunging abyss of incomprehension.

But Guérin goes still further. We must abandon the simplistic idea, he says, that intellectual progress forges ahead on a single track, and that the most advanced species necessarily maintain their contact with and understanding of the lower species. Human experience shows the contrary. A grafted branch is cut off and isolated from its own natural resources, yet it thrives; but the enormous differences in psychic and sensorial equipment that separate animal species estrange them from each other.

Opposing Guérin, scientists like Hoyle, Morrison and Drake agree on the probability of a single direction that all intellectual development must take. If true, this would simplify the mechanics of the communications problem. In the view of these astronomers and physicists, life's paths are more strictly programmed than anyone has ever imagined. Life cannot make any choice other than the chemistry of carbon, and it will only tolerate physical conditions similar to those which govern the Earth and which alone can ensure the interplay of their metabolism.

By extension, this restrictive, authoritarian direction must also impose its constraints on evolving intelligence. Behind the diversity of terrestrial forms, Hoyle espies a powerful unifying force which, in the higher species, implies a sensorial and cognitive equipment of the same kind.

In support of this principle, Morrison cites the phenomenon of convergencies, familiar to zoologists, which confirms his theory that nature adopts like solutions when she is called upon to solve like problems. In the deep sea, she created a mammal, a fish and a prehistoric reptile in much the same way to make them all fast swimmers; and despite the wide disparity in their structures, the ichthyosaurus, the shark and the dolphin bear a remarkable resemblance.

Another example: the bird, the bat and the flying batrachian of the primary era developed quite different body structures to sprout their wings. Although frames and matter are dissimilar, nevertheless life is oriented to the same purpose.

Another example of convergency is indeed impressive: life has re-invented the eye three times—in the vertebrates, of course; in the various families of arthropods, crustaceans and insects, all endowed with extremely elaborate ocular systems; and in the mollusc. The eyes of an octopus are not much different from ours.

In his affable way, Fred Hoyle tells us that the eye is inevitable in terrestrial physical conditions, just as is the brain. "In my opinion", he once said, "all creatures of high intelligence living on planets like ours and having the same basic system of chemistry are no doubt animals much like ourselves. They probably bear important anatomical similarities—skeleton, head, eyes and so forth." He thinks that "the technological leap forward that brought us out of the stone age up to a refined culture and technology, couldn't have happened simply by chance. The history of our species falls into a determined plan. We are advancing along an inevitable route."

But life on Earth provides us little opportunity to experiment on intelligent exchanges with beings of a totally different development.

Once at a Green Bank conference, the neurologist John C. Lilly brought up the prospects of communications between man and the dolphin, a study in which he specializes.

The dolphin's brain, Lilly reminded his audience, is highly developed. Its dimensions, structure and quality of nervous tissue suggest that this creature might well command a relatively sophisticated language. All social animals employ signals to manifest repulsion or attraction. Some of them even possess signalling systems that transmit all sorts of information to others of their species. Ants emit "words" together with various odors, depending upon what they want to say. Everyone knows about the dance of the bees. The song of some birds juxtaposes phrases with simple signals. Here we can perceive a distant suspicion of an articulated language. Our own language is built on "representation." What was once a grunt has become an abstraction. With man, or rather with the infant, the language that comes to him depends largely upon the use he has made of his eyes and hands to identify the objects around him and establish his topological relations with the world.

Apart from his exceptional brain, the dolphin is also blessed with a delicately refined sensorial apparatus. Because of his sonar superiority, he practises echo-localization with astounding precision. His acute hearing apparently registers all sorts of holograms representing the objects detected—three-dimension images so exact that it is hard to catch him out. Considering his intelligence, his powers of detection and the wide range of his acoustical signals (dolphins virtually carry on conversations among themselves), Lilly had high hopes that the creature commanded an evoluted language which could express mental representations.

Although he pointed out that the dolphin's brain might function in ways radically different from ours—to study it would be a long and difficult task—he did not discount the possibility of acquiring a scientific understanding of his language and his psychic system. But he threw in the towel. All efforts to decode the language of dolphins failed. In vain have experimenters analyzed their cries by utilizing techniques closely akin to those Project Cyclops intends to mobilize in its search for galactic signals. The recordings of their whistles stubbornly resisted "Fourrier transformations," a mathematical method of singling out coherent signals superimposed on an anomalous emission. Lilly, who carried his studies to great lengths, chides us for extrapolating the conduct of these mammals from our own patterns of behavior and then becoming duped by the mimicry we have taught them.

During his experiments, dolphins were trained to recognize fairly complicated signals conveyed to them in modulated whistles, a language that includes up to 256 words of four syllables, multiples of binary signals easily combined by computers. The U.S. Navy, which actually enlists dolphins for military duty, uses the machine-animal system with remarkable efficiency. But such thralldom to electronics in man's relations with animals disheartened Lilly.

While dolphins will respond to orders issued by artificial signals, their specific means of communicating among themselves remains a total mystery to us. Yet, they speak their own language to exchange information with other species, including the whale.

For the present at least, Guérin's arguments stand. Just as we cannot communicate to any great extent with other animals, he believes that the mental and technological chasm which must segregate the galactic societies rules out any communications between equals. In any event, in practice "a mutual exchange of information at a distance between them and us on the basis of our techniques" is most improbable. It is far more reasonable, he thinks, "to suppose that it is on their level and on the basis of their techniques that they have worked out adequate means—if they wish to know us—to study scientific communities in their infancy, such as ours. But if this is true, then their devices are, for the time being, beyond our scientific comprehension, if not our capacity for detection."

Like McDonald, in his allusions to UFOs, Guérin believes that the "saucer phenomenon" could be attributed to extra-terrestrials who are investigating us earthlings, commanding powers unknown to us and likewise beyond our grasp. It is possible that outer-spacemen are putting us to a close, detailed scrutiny, much as we study fruit flies and drops of water under the microscope.

Oppenheimer, too, wondered whether the science of an

outer space civilization would be intelligible to us. He had in mind the *uncertainty principle* to which all nuclear physicists must accommodate themselves and which prevent them from determining simultaneously both the position and the velocity of a particle. They cannot localize it when they know its speed, and vice versa.

Oppenheimer was convinced that the world's science is hindered in every respect by limitations of this kind. Without knowing it, when we choose to answer certain questions, we close our minds to other possible lines of interrogation. Perhaps on some distant planet, fundamental questions are posed differently, perhaps galactic communities are all isolated, each one confined within its own sphere of knowledge.

Since the turn of the century, man has learned that he can grasp reality by more than one single method. If societies in outer space have evolved their science along different directions, it will be all the harder to understand them. All the same, Oppenheimer contended that mutual understanding is not unthinkable. Far from chaining us to our perch, this reflection on the limits and precariousness of human observation provided us the springboard for a take-off into better world understanding; and when the human horizon will have expanded, we shall then be able to integrate our limited information and relative knowledge into more enlightened systems. At the same time, beings far ahead of us should be able to appreciate our retarded mental condition and recognize it as a past stage in their own scientific evolution.

Von Hoerner has put his finger on another cardinal uncertainty: do galactic societies *want* to make contact with each other? This would depend upon the age of each one, its level of development and its dynamics. Certainly the question applies to us, and we must decide whether we care about joining that supercivilization of galactic communications that Hoyle so pleasantly evokes, declaring it to be the greatest prospect of expansion open to mankind.

Von Hoerner continues: if earthlings are unwilling to

make the necessary efforts to join with other intelligent worlds, which more than likely exist, this would signify that we are resigned to languishing into an era of organized stagnation, and such a course could threaten to inflict the final solution on human history. Predominantly scientific and technological civilizations apparently do not endure for long, otherwise we should have found them. Technical progress harbors mortal dangers. When societies survive their explosive development, it is probable that they adopt other priorities and lose all interest in their soaring scientific and technological achievements. Von Hoerner dwells insistently on this resignation of a society "in danger of progress," which prefers technical utilitarianism to great scientific adventure and retreats into a sort of Malthusianism of the intelligence: survival by withdrawal.

In the other prospect, certainly the same society will exploit its brief era of techno-scientific dynamism to mobilize its intellectual and technical resources toward gaining access to the galactic community. This escalation is held to be a rare opportunity for a planetary race to survive and forge ahead.

We have seen that the American researchers von Hoerner, Drake and Sagan, have attempted to demonstrate the superiority of telecommunications over expansion in space. Von Hoerner in particular shows the futility of any attempted stellar colonization which could not in any way provide a solution to our problem of overpopulation. He has estimated that, by clinging to the hypothesis of a total mastery of space travel at the ideal speed of light, humanity would have exhausted its resources of colonization within five hundred years without realizing any notable improvement in its situation.

Hoyle thinks that the future of man lies in communications. His power can only grow if he finds access to sources of momentous information as yet untapped. Here the essential factor is to plug into other worlds' channels, and widen the circle of the intelligent community. The level of a civilization will be measured by the share of knowledge it acquires.

Along with a number of other astrophysicists, Hoyle and Sagan are convinced that electromagnetic rays offer a realm of study and exploitation that must be common to the advanced societies in space. When we set out in this direction, we are obliged to admit that the spectrum of radio waves provides the most likely meeting ground.

A relative uncertainty in the matter of contact techniques, of course, persist. But the hertzian waves remain the most acceptable of all the candidate methods we know about. But then, we have no other alternative than to make the most of the resources offered us by the best technology we have.

BIBLIOGRAPHY

Biraud, François and Jean-Claude Ribes. *Le dossier des civilisations extraterrestres*. Paris: Fayard, 1970.

Bulletin of the Atomic Scientists (The Condon Report and U.F.O.), Vol. XXV, Numbers 3 and 10. Chicago: 1969.

Guérin, Pierre. *Planètes et satellites*. Larousse, 1967.

Hearings Before the Committee on Science and Astronautics, House of Representatives: Symposium on U.F.O.s 1968. Springfield, Va.: Clearing House.

Hoyle, Fred. *Of Men and Galaxies*. Seattle: University of Washington Press, 1964.

Jung, C. G. *Un mythe moderne (Ein moderner Mythus)*. Paris: Gallimard, 1961.

McDonald, James E. "Objets volants non identifiés, le plus grand problème scientifique de notre temps?" translated by René Fouéré, *Phénomènes spatiaux* (Paris, 1969).

Oliver, B. M. "State of the Art in the Detection of Intelligent Extraterrestrial Signals," *International Astronautics Federation Conference* (Vienna, 1972).

Sagan, Carl, ed. *Communication with Extraterrestrial Intelligence*, Proceedings of a conference held at the Byurakan Astrophysical Observatory, U.S.S.R. Cambridge, Mass: MIT Press, 1973.

Sagan, Carl. *The Cosmic Connection*. New York: Anchor Press, Doubleday, 1973.

Shklovski, I. S. *Univers—Vie—Raison*. Moscou: Editions de la Paix, 1962.

Shklovski, I. S. and Carl Sagan. *Intelligent Life in the Universe*. San Francisco: Holden-Day, 1966.

Soustelle, Jacques. *Les quatre soleils*. Paris: Plon, 1967.

Sullivan, Walter. *We Are Not Alone*, revised edition. Harmondsworth: Penguin Books, 1970.

von Hoerner, Sebastian. *Population Explosion and Interstellar Expansion*. Göttingen: Vandenhoek and Rupprecht (to be published).

———. "Astronomical Aspects of Interstellar Communication," *Acta Astronautica* (to be published).